150 CAPTIVATING CHEMISTRY EXPERIMENTS USING HOUSEHOLD SUBSTANCES

By BRIAN ROHRIG

Other Books by Brian Rohrig:
150 More Captivating Chemistry Experiments Using Household Substances
101 Intriguing Labs, Projects, and Activities for the Chemistry Classroom
39 Amazing Experiments with the Mega-Magnet
39 Fantastic Experiments with the Fizz-Keeper
39 Dazzling Experiments with Dry Ice
39 Spectacular Experiments with Soda Pop

All books can be ordered directly from FizzBang Science

The photos included in this text are not intended to imply a commercial endorsement of any product.

All experiments in this book are to be performed only under competent adult supervision. Neither the author nor the publisher assume any liability whatsoever for any damage caused or injury sustained while performing the experiments contained in this book.

12 11 10 9 8 7 6 5 4

Published by FizzBang Science
807 Murlay Drive
Plain City, Ohio 43064
www.fizzbangscience.com
info@fizzbangscience.com

ISBN 0-9718480-2-5

For Brandon

Acknowledgements:
Thanks to Frank Reuter for carefully and thoroughly editing this text. Any and all errors are strictly my own.

Cover design by Kristi Gerner.

Line drawings by Lydia Cooper.

Books printed by Network Printers in Milwaukee, Wisconsin.

TABLE OF CONTENTS

CHAPTER 4: GASES, LIQUIDS, AND SOLIDS

CHAPTER 5: CHEMICAL REACTIONS

CHAPTER 8: SOLUTIONS AND SOLUBILITY

CHAPTER 9: POLYMERS

CHAPTER 10: LIGHT

INTRODUCTION

This book is for anyone who loves science – students, parents, or teachers. The 150 experiments contained in this manual are fully explained and illustrated. Each experiment can be performed with ordinary household substances that are easily obtained from the grocery or hardware store. A chemistry lab is not required to do any of these experiments. No expensive glassware or obscure chemicals are needed. You will be amazed at how much you can accomplish on a limited budget.

These experiments are suitable for grades K-12. I have done many of these activities with high school students, as well as with my own children when they were very young. It is never to early to instill a love for science in our children.

A student with a curiosity for science will find this book invaluable, since all of the experiments can be done right at home with materials found around the house. Adult supervision is a must, however, and appropriate safety precautions are given with each experiment.

Teachers will find this book to be a wonderful asset to their science curriculum. Since many teachers are on a limited budget, this book will provide many activities at minimal cost. Science teachers at all levels can make use of this manual, regardless of their training or background.

The homeschooling movement is growing by leaps and bounds. If you teach your children at home, this book will be an excellent resource.

To get the most out of every experiment, read the instructions carefully, follow all safety precautions, and make careful observations. But above all, have fun!

Brian Rohrig

SAFETY PRECAUTIONS

Please read thoroughly before proceeding with the experiments in this book.

1. All experiments are to be done only with competent adult supervision.

2. Never taste or drink the product of any reaction, unless specifically instructed to do so.

3. Never inhale any chemical substance. If you must smell something, wave your hand over the opening of the container and gently waft fumes towards your nose.

4. Always wear safety goggles when doing any experiment.

5. Always have a fire extinguisher nearby when using open flames.

6. Never add water to an acid. Always add acids to water.

7. If you spill a chemical on your skin, immediately wash it off with copious amounts of water.

8. If a chemical splashes in your eyes, rinse eyes thoroughly under water for 15 minutes and seek outside medical help immediately.

9. Always read the label of any chemical thoroughly before using.

10. Dispose of any chemical only according to the instructions on the label.

11. Store all chemicals out of the reach of children.

12. If smoke or fumes are produced, only perform the experiment under a fume hood or outside.

13. Never heat a closed container.

14. Keep all flammables away from open flame.

15. If you have long hair, tie it back when working around flames.

16. Do not wear loose-fitting or baggy clothing when working with flames or chemicals.

17. Wash hands thoroughly when finished.

CHAPTER 1 DENSITY

Density is defined as the amount of space a certain amount of matter occupies. Since 1 gram of water occupies a volume of 1 milliliter, water has a density of 1 g/mL.

Density can also be defined as a measure of how far apart the molecules are within a substance. Substances like gases have a very low density because their molecules are very far apart. Solids have a higher density because their molecules are closer together.

Mass refers to how much matter exists within an object. Matter is composed of atoms, which are the building blocks of all substances. The basic metric unit of mass is the gram.

Volume refers to how much space an object occupies. The bigger an object is, the greater its volume. The basic metric unit of volume is the Liter.

Buoyant force refers to the force that liquids or gases exert on other objects. The buoyant force always acts upward, and is responsible for keeping objects afloat in water, or making things rise in the air, like a helium balloon.

In this chapter, we will explore the fascinating world of density.

Since density affects so many natural phenomena, it is an excellent place to begin our study of chemistry . . .

Density – Experiment # 1:
DOES HOT WATER RISE?

Objective: To demonstrate that hot water is less dense than cold water.

Materials:

- Two 16 oz glass Snapple bottles (or equivalent)
- Hot plate or stove
- Pan or beaker
- 3 x 5 inch index card
- Food coloring

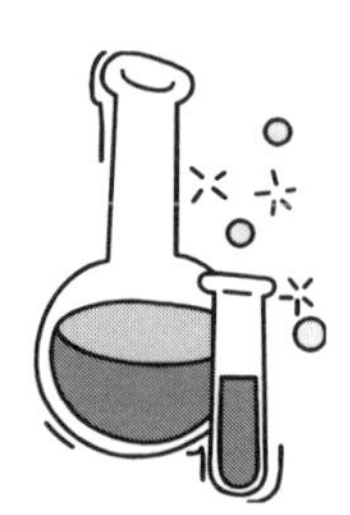

Safety Precautions: Be careful with boiling water. Use a potholder to handle the bottle after filling it with hot water.

Procedure:

1. Heat up approximately 500 mL of water to near boiling.
2. Pour this hot water into a Snapple bottle, up to the brim.
3. Add a few drops of food coloring.
4. Pour cold water into the other Snapple bottle, up to the brim. Do not add food coloring to this bottle.
5. Place a 3 x 5 card over the top of the bottle of cold water.
6. Carefully invert the bottle of cold water, with the card over its opening, so that it is resting on top of the bottle of colored hot water.
7. Carefully slip the card from between the two bottles, being careful not to move either bottle. This will allow the hot and cold water to mix. Observe for several minutes.

Explanation: The cold water sinks, since it is denser than the hot water. As the cold water sinks, the hot water is displaced and forced upward. This clearly shows that hot water is less dense than cold water. It is important to note that the hot water does not actually rise. It only appears to rise because the cold water is pushing it upward.

Fluids become less dense when heated because heating causes their molecules to move faster. As the molecules move faster, they spread farther apart from one another.

If the above system is allowed to remain at rest for several minutes, both bottles will be the same color and the same temperature. We can then say that equilibrium has been established.

This experiment also demonstrates convection, which is the transfer of heat by the movement of a fluid. A convection current is established in the bottles, since hot water is being forced upward by the denser cold water. This movement of hot water transfers heat between the two bottles. A convection current will continue to operate in the bottles until equilibrium has been established.

Repeat the above procedure, except this time place the bottle of hot colored water on top of the cold water. What do you think will happen?

Density – Experiment # 2:
SINK or FLOAT?

Objective: To discover why diet soda is less dense than regular soda.

Materials:

- 10 gallon aquarium or other large transparent container
- Cans of diet and regular soda
- Double pan balance (optional)

Safety Precautions: None

Procedure:

1. Fill the aquarium halfway with water.
2. Predict what will happen if an unopened can of regular soda is placed in the aquarium. Drop a can into the aquarium.
3. Predict what will happen if an unopened can of diet soda is placed in the aquarium. Drop a can into the aquarium.
4. Place the cans of diet and regular soda on a double pan balance. Observe.
5. Drop in other cans to see what will happen. Try unopened cans of other beverages that come in aluminum cans.

Explanation: This fascinating demonstration is an excellent way to learn about density. We are all familiar with the basic concept of sinking and floating. Empty cans float, rocks sink. Objects less dense than water float, and those denser than water sink.

If both cans are put on a double pan balance, it is clear that the regular soda is heavier than the diet soda. This demonstrates the difference between mass and volume. Mass refers to how much stuff (how many protons and neutrons) exist within an object. If something is heavier than another object, it contains more mass, or simply contains more protons and neutrons. Mass is measured in grams.

Volume, on the other hand, refers to how much space an object occupies. For fluids, volume is usually measured in liters (L) or milliliters (mL). There are 1000 mL in one Liter. Since both cans have the same volume, the heavier can must have a greater mass. We can thus conclude that the heavier can is denser than the lighter can.

Diet sodas usually contain aspartame, an artificial sweetener, while regular sodas use sugar. Reading the label on a can of sugared soda will reveal that a single can may contain over 40 grams of sugar. If this quantity is massed out on a balance, it is an impressive amount of sugar. Artificial sweeteners such as aspartame can be up to 200 times sweeter than sugar. Therefore, only a tiny amount of aspartame is needed.

Both sugar and aspartame are denser than water, which can be easily demonstrated by adding small amounts of each to a container of water. Both powders sink if placed in water. It is the amount of each sweetener used that is crucial. The 40 grams or so of sugar added to a can of regular soda makes it sink. The relatively tiny amount of aspartame used in diet sodas will have a negligible effect on the mass, enabling the can to still float.

If you add 40 grams of sugar to a nearly full can of water that is barely floating, the can will sink. If a gram of aspartame is added to a nearly full can of water that is barely floating, it will probably continue to float.

So why do cans of diet soda float? There is a little bit of space, termed head space, above the fluid in each can of soda. This space is filled with gas, which is much less dense than the soda itself. It is this space above the soda that lowers the density of cans of diet drinks just enough to make them float. Sugared drinks still have this head space, but the excessive amount of sugar added makes the can denser than water. You can occasionally find a can of sugared soda that floats. Can you provide a hypothesis to explain this behavior?

Related Experiments: Try placing different objects in the aquarium to see whether they sink or float. Predict their behavior beforehand. Some interesting objects to try are: candles, fruit, vegetables, wood, plastic, and aluminum foil. If you have a piece of pumice on hand, it will float. Pumice, a porous rock that is the product of volcanic eruptions, is one of the few rocks that floats! There is also a type of wood, known as lignum vitae, that actually sinks. It has a density of 1.39 g/mL, which is greater than water. A sample of this wood can be ordered from Museum Products at 800-395-5400 or www.museumproducts.com.

Density – Experiment # 3:
BOBBING RAISINS

Objective: To discover why raisins that are placed in a carbonated beverage will rise up and down for several minutes.

Materials:
- Transparent plastic cups
- Raisins
- Clear carbonated beverage, such as Sprite, 7 Up, or Mountain Dew

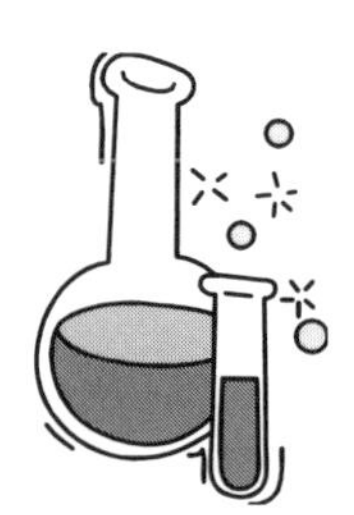

Safety Precautions: None

Procedure:
1. Pour a glass of clear, carbonated beverage. (Alternately, you may keep the beverage in the bottle itself.)
2. Add several raisins. Observe.

Explanation: The raisins will bob up and down for several minutes. This phenomenon is fascinating to observe. Close observation will reveal why this is happening. Since the surface of the raisins are so rough, tiny bubbles of CO_2 will be attracted to the raisin, attaching to its surface. These bubbles will increase the volume of the raisin substantially, but contribute only very little to its mass. As a result, the overall density of the raisin is lowered, causing it to be buoyed upward by the more dense fluid surrounding it.

Archimedes' Principle states that the buoyant force exerted on an object surrounded by a fluid is equal to the weight of fluid displaced by that object. Since the raisins now have a greater volume, they displace more water, causing the fluid to exert a greater buoyant force. The buoyant force is what pushes the raisins to the top.

Once the raisins reach the air at the surface, the bubbles pop. This makes the raisins denser, causing them to sink. As more bubbles adhere to the raisins, they again become less dense and are pushed back up by the fluid. This experiment very clearly shows that an increase in volume (as long as the mass increase is negligible) will lead to a decrease in density. The bubbles that attach to the raisins can be thought of as little life jackets, which make the raisins more buoyant by increasing their volume.

If the raisins are added directly to the bottle, replace the cap on the bottle after they have made a few cycles. After a few minutes, the raisins will all stay at the bottom. This is because CO_2 gas is prevented from leaving the bottle. As a result, pressure builds up in the space above the fluid. This pressure is transmitted throughout the fluid, the result being that the bubbles cannot grow as large. Therefore, the volume of the raisins does not increase enough to lower their density to the point where they are less dense than the fluid. When the cap is removed, the pressure above the raisins is decreased, allowing the bubbles to grow larger, causing them to resume the cycle of bobbing up and down.

Experiment with different types of carbonated drinks to see which works the best. Instead of raisins, try other substances such as grapes, mothballs, spaghetti, and Grape Nuts cereal.

Density – Experiment # 4:

THE EXPANDING NIPPLE

Objective: To discover how much carbonation exists in a can of soda.

Materials:

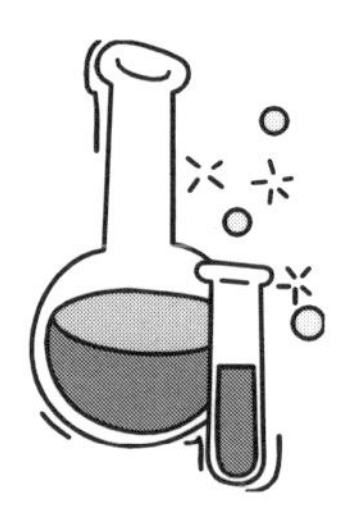

- Can of soda
- Baby bottle
- Blind nipple (available from drugstore or free from Evenflo by calling 800-356-BABY)

Safety Precautions: Make sure the bottle is sealed tightly before shaking. Replace the blind nipple with a new one after doing this experiment a few times. Wear safety goggles – occasionally a blind nipple will burst while doing this experiment.

Procedure:

1. Fill a baby bottle nearly to the top with a freshly opened can of soda.
2. Screw on the cap so it is very tight, with the blind nipple attached.
3. Shake vigorously. What happens to the nipple?
4. Turn the bottle upside down to fill the nipple with the soda. This will show how much gas is in the soda.
5. Turn the bottle over again and unscrew the cap to release the CO_2 gas. If shaken up again, what happens?

Explanation: This classic experiment demonstrates that there is a great deal of carbonation in a can of soda. It also illustrates that air can act as a nucleation site. By introducing air into the soda by shaking, the molecules of carbon dioxide will adhere to the molecules of nitrogen and oxygen in the air, growing into large bubbles that escape from the solution. These molecules of gas exert so much pressure that they cause the nipple to expand. These bubbles are forced to the top of the solution and collect in the nipple, because they are much less dense than the liquid. When the gas is released, the nipple goes back to normal size, and will not expand again because the soda has gone flat.

Density – Experiment # 5:

IS SALT WATER DENSER THAN PURE WATER?

Objective: To demonstrate the difference in density between fresh and salt water.

Materials:

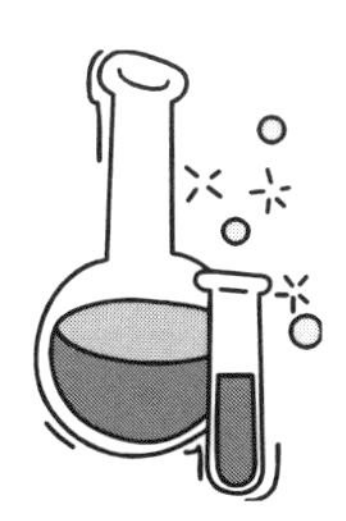

- Egg
- Salt
- Beaker or drinking glass

Safety Precautions: None

Procedure:

1. Place an egg in a glass of water. The egg should sink. (If the egg floats, try another egg. When eggs get old, they fill with gas and may float.)
2. Remove the egg and add several teaspoons of salt. Stir until most of the salt has dissolved.
3. Place the egg in the salt water. It should float. If it does not, continue adding more salt.

Explanation: This is a classic experiment that clearly demonstrates that salt water is denser than pure water. Objects floats because a liquid is exerting a buoyant force on the floating object. If the buoyant force acting on an object is greater than the force of gravity acting on that object, then that object will float. The denser the liquid, the greater the buoyant force it can exert. Salt water, being denser than pure water, can exert a greater buoyant force, enough to cause the egg to float. This explains why it is much easier for people to float in salt water than in fresh water. In some salty lakes, such as the Dead Sea, it is nearly impossible to sink!

Density – Experiment # 6:

MAKING A DENSITY COLUMN

Objective: To demonstrate that some liquids are more dense than others.

Materials:

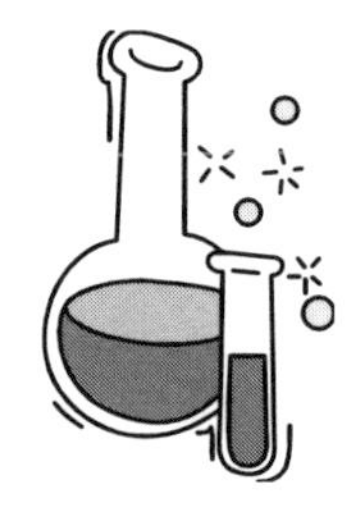

- Karo syrup (dark)
- Vegetable oil
- Food coloring
- 20 oz transparent soda bottle

Safety Precautions: None

Procedure:

1. Pour some dark Karo syrup into the bottle, to a depth of several centimeters.
2. Add some colored water to the bottle.
3. Add some vegetable oil.
4. Replace the cap and keep the bottle as a permanent display. Do not shake or stir.

Explanation: The three liquids you mixed form distinct layers due to differences in density, as well as solubility. The more dense liquids will sink to the bottom, and the less dense liquids will float to the top. This illustrates why oil always floats on water.

Experiment with other liquids to see how many layers you can form. Do not mix cleaning fluids or flammables, and always read the labels before mixing any two substances. By using items found in your kitchen, you should be able to obtain several more layers. You can also drop small solids into your column, to see where they will fall. Pieces of rubber, plastic, and wood should all float at different levels within your bottle due to density differences.

Density – Experiment # 7:

MAKING A CARTESIAN DIVER

Objective: To demonstrate positive, negative, and neutral buoyancy.

Materials:

- 2-Liter plastic bottle
- Glass eyedropper with rubber bulb

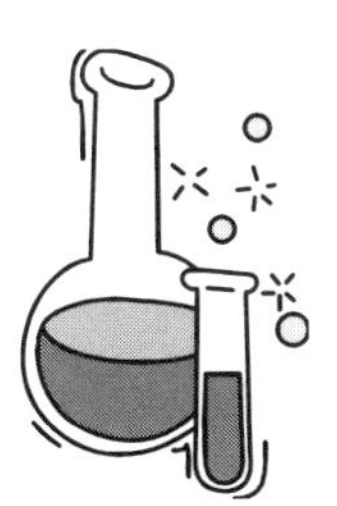

Safety Precautions: None

Procedure:

1. Fill a 2-L bottle nearly to the brim with water.
2. Drop in the eyedropper.
3. Tightly cap the bottle.
4. Squeeze the sides of the bottle until the eyedropper sinks to the bottom.
5. Attempt to keep the eyedropper suspended by regulating the pressure on the bottle with your hands.

Explanation: By squeezing the bottle, you force water up into the eyedropper, making it denser than water. Therefore the eyedropper sinks to the bottom. When an object sinks, it is negatively buoyant.

When the pressure on the bottle is released, the compressed air inside the eyedropper forces the water back out, causing it to be less dense than water. This causes the eyedropper to float back to the top. When an object floats, it is positively buoyant.

By regulating the amount of water that is forced into the eyedropper, it can be suspended in one place, neither sinking nor floating. This is known as neutral buoyancy.

This experiment also provides a good illustration of Boyle's Law, which states that putting pressure on a gas causes its volume to decrease. As water is forced into the eyedropper, the air inside has no place to go, so it is simply compressed in the eyedropper. Since the same amount of air is taking up less space, it is under greater pressure.

Density – Experiment # 8:

HOW TO MAKE A HEAVY BALLOON

Objective: To demonstrate that carbon dioxide is denser than air.

Materials:

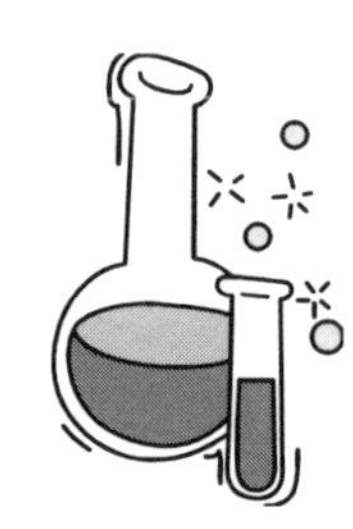

- 20 oz plastic soft drink bottle
- Balloons
- Baking soda (sodium bicarbonate – $NaHCO_3$)
- Vinegar (dilute acetic acid – $HC_2H_3O_2$)
- Funnel

Safety Precautions: Wear safety goggles.

Procedure:

1. Add 3 teaspoons of baking soda to a soft drink bottle.
2. Using a funnel, fill a balloon with vinegar until it is nearly filled to the top.
3. Stretch the opening of the balloon over the mouth of the bottle while pinching above the main body of the balloon to prevent the vinegar from escaping.
4. Once the balloon is firmly fastened over the mouth of the bottle, invert the balloon to allow the vinegar to fall into the bottle.
5. When the balloon ceases to expand, pinch the balloon, remove it, and tie it off.
6. Blow up another balloon with your breath to the same size as the carbon dioxide-filled balloon. Tie off this balloon.
7. Drop both balloons from the same height and note which one hits the ground first.

Explanation: The baking soda and vinegar react to form carbon dioxide gas, which fills the balloon. The balanced chemical equation is as follows:

$$\underset{\text{vinegar}}{HC_2H_3O_{2(aq)}} + \underset{\text{baking soda}}{NaHCO_{3(aq)}} \Rightarrow \underset{\text{sodium acetate}}{NaC_2H_3O_{2(aq)}} + \underset{\text{water}}{H_2O_{(l)}} + \underset{\text{carbon dioxide}}{CO_{2(g)}}$$

Since carbon dioxide is about three times denser than air, the CO_2-filled balloon will fall to the ground at a much faster rate than the air-filled balloon.

Related Experiment: Allow both balloons to remain inflated side by side overnight. Do you notice a difference in the rate at which each loses air? Can you provide an explanation?

Density – Experiment # 9:
FUN WITH CARBON DIOXIDE

Objective: To demonstrate how carbon dioxide can be used to extinguish a flame.

Materials:

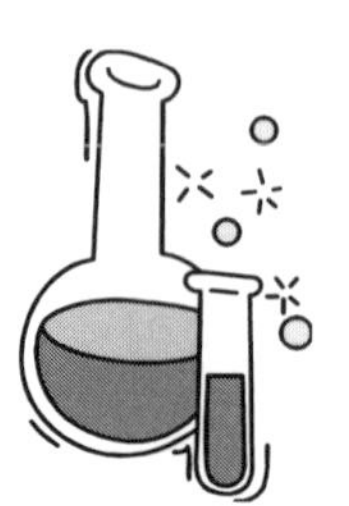

- Glass bottle (a Snapple bottle is ideal)
- Baking soda
- Vinegar
- Candle
- Matches

Safety Precautions: Wear safety goggles. Exercise caution when using flames.

Procedure:

1. Light a candle and set it aside.
2. Place 3 teaspoons of baking soda in a bottle.
3. Add about 30 mL of vinegar to the bottle.
4. After the bubbling has ceased, attempt to pour the invisible gas that is now in the bottle over the candle flame. Be careful not to pour any liquid over the flame. Observe what happens to the flame.

Explanation: Carbon dioxide gas is formed as a result of a chemical reaction between the baking soda and vinegar (see the previous experiment). CO_2 gas can actually be poured, since it is denser than air. When poured, it always sinks downward. As a result, any oxygen around the flame will be displaced by the CO_2 gas. This causes the flame to be extinguished.

Since CO_2 is so dense, it tends to remain in the bottom of its container for several minutes. If a lit match is inserted into the container, it will be extinguished. CO_2 gas can even be poured from one container to another, and then poured over a flame to extinguish it.

Density – Experiment # 10:
NEUTRAL BUOYANCY: PART 1

Objective: To make a film canister neutrally buoyant in water.

Materials:
- Film canister
- Assorted small nails, bolts, paper clips, etc.
- Eyedropper
- Transparent glass or beaker
- Salt

Safety Precautions: None

Procedure:
1. Place assorted objects in the canister until it just floats.
2. Now add water with the eyedropper one drop at a time until the canister remains suspended in the water below the surface. It should neither sink nor float, with no portion of the canister touching the bottom or sticking out above the surface of the water.
3. Attempt to adjust the density of the water in the beaker by adding salt. Does this make it easier to achieve neutral buoyancy?

Explanation: This experiment can be extremely frustrating, but with a great deal of patience, neutral buoyancy can be achieved. An object is neutrally buoyant if its density is exactly the same as that of its surroundings. Therefore a neutrally buoyant object neither sinks nor floats. If an object is less dense than its surroundings, it will float. Floating objects exhibit positive buoyancy. If an object is denser than its surroundings, it sinks. Sinking objects exhibit negative buoyancy.

Probably the easiest way to achieve neutral buoyancy is by adding enough mass to the canister so that it barely sinks. Then add salt a little at a time to the beaker of water. This will increase the density of the water just enough to make the canister rise. Since more dense fluids exert a greater buoyant force, the canister will be pushed upward.

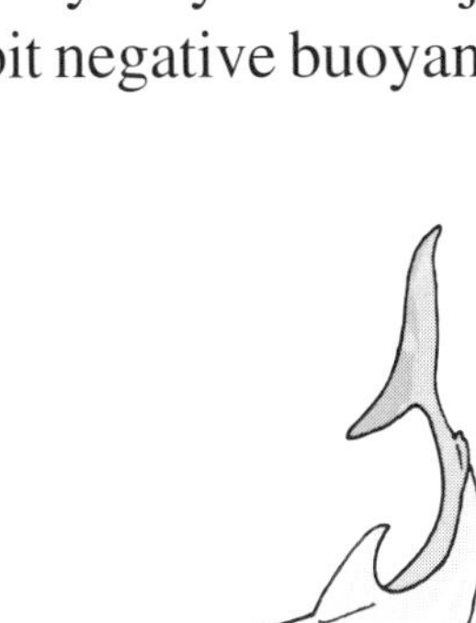

Fish that stay in one place underwater are neutrally buoyant. They can adjust their density by taking in or releasing air from their swim bladder. As a result, fish can conserve energy by remaining motionless underwater. This is only possible by attaining a state of neutral buoyancy. Sharks, on the other hand, lack a swim bladder. Their huge liver stores a large amount of oil, which makes them more buoyant. But they are still denser than water, so they must swim continually or else they will sink. Submarines are also neutrally buoyant if they remain in one place underwater without touching the bottom.

Related Experiment: Many soft drinks that come in aluminum cans are very close to neutral buoyancy. Try to achieve neutral buoyancy with one of these cans by adding salt to the water.

Density – Experiment # 11:

NEUTRAL BUOYANCY: PART 2

Objective: To make a helium balloon neutrally buoyant in air.

Materials:

- Helium-filled balloon
- Twine
- Styrofoam cup
- Eyedropper

Safety Precautions: None

Procedure:

1. Using a pencil, punch two holes near the top of a Styrofoam cup on opposite sides.
2. Run a piece of twine about 30 cm long through both holes of the cup and then tie the twine in a knot so it can support the cup.
3. Tie the other end of the twine to a helium-filled balloon. If the balloon does not rise at this point, you will either need a bigger balloon or a smaller cup. You can also try tearing off pieces of Styrofoam from the cup to make it lighter.
4. Now add drops of water to the cup with the eyedropper until the balloon remains suspended in midair. It helps if air currents in the room can be minimized as much as possible.

Explanation: Like the last experiment, this one requires a great deal of patience. Achieving neutral buoyancy can be a rather frustrating experience! Even tiny changes in density can determine whether an object will rise or fall.

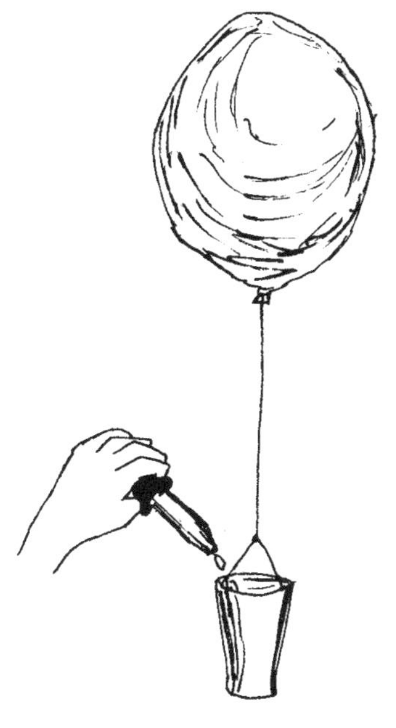

A helium balloon rises because it is less dense than air. The air exerts a buoyant force large enough to make the balloon rise. If the buoyant force of the air pushing upward is greater than the force of gravity pulling downward (also known as weight) then the balloon will rise. If the buoyant force of the air is less than the force of gravity, then the balloon will sink. If the buoyant force of the surrounding air is equal to the force of gravity acting on the balloon, then the balloon will be neutrally buoyant. By adding water to the cup, you are changing the density of the balloon-cup combination until it is equal to that of the surrounding air.

Dirigibles, such as the Goodyear Blimp, are in a state of neutral buoyancy anytime they remain suspended in midair without rising or falling. Their density is computer-controlled to enable them to maintain a state of neutral buoyancy in the air.

CHAPTER 2
AIR PRESSURE

On the earth's surface, you are at the bottom of an ocean of air. This ocean of air is deepest at sea level, and gradually thins out into outer space several hundred miles up. The higher your altitude the less air pressure you experience, because the weight of the air above you is less.

At sea level, air pressure is exerting a pressure of 14.7 pounds per square inch (psi). This is sometimes referred to as simply 1 atmosphere (atm).

In this chapter, we will explore the awesome power of this invisible air pressure that constantly envelops us . . .

Air Pressure – Experiment # 1:
THE IMPLODING SODA CAN

Objective: To crush a soda can with air pressure.

Materials:

- Empty soda can (an *Orange Crush* can is particularly fitting!)
- Stove or hot plate
- Shallow metal pan
- Tongs
- Ice

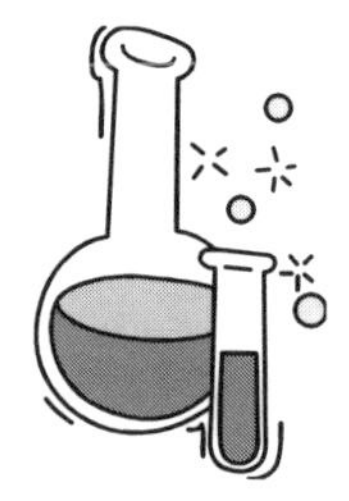

Safety Precautions: Wear safety goggles. Exercise caution with boiling water. Do not grasp soda can without tongs.

Procedure:

1. Fill an empty soda can to a depth of approximately two cm with water.
2. Place the can on a hot plate and heat the water within the can to boiling.
3. Fill a shallow pie pan with water and add ice cubes.
4. Continue heating until condensed steam rises from the can for about 1 minute.
5. Grasping the can with tongs, quickly invert the can and plunge it into the pan of cold water. Observe.

Explanation: As the soda can is heated, the air in the can is also heated, expanding and rising out of the can. The can then fills with steam, and this steam pushes out the rest of the air from the can. When the can is inverted into the pan of cold water, the steam immediately condenses back into liquid water. This can "full" of steam will condense into only a few drops of water, forming a nearly complete vacuum within the can. Since the opening of the can is under water, air cannot rush in to fill this vacuum. However, the outside of the can is still surrounded by air, and air pressure pushes down with a force of 14.7 pounds per square inch (psi). Normally, air on the inside of a can will push outward at the same pressure and balance out this outside air pressure. However, since we have removed the air from inside the can, and the outside air pressure is still acting on the can, the can is instantly crushed by this outside air pressure. This experiment provides an excellent example of the tremendous power of the air around us.

Air Pressure – Experiment # 2:

THE POWER OF THE AIR

Objective: To support a cup of water using air pressure.

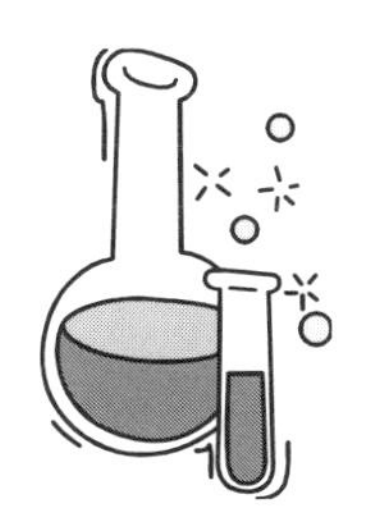

Materials:

- 8 oz paper cup
- 3 x 5 inch index card

Safety Precautions: Do this over a sink in case of an accident!

Procedure:

1. Fill your cup to the brim with water.
2. Place a 3 x 5 card over the top of the cup. Make sure that the card covers the top of the cup completely. If not, substitute with a larger card or a smaller cup.
3. Holding your hand over the card, slowly invert the cup of water.
4. Remove your hand, and the card appears to be supporting the entire cup of water!

Explanation: Air pressure is being exerted in all directions with a force of 14.7 lbs per square inch. Since the card has a total area of 15 square inches, air pressure is exerting a force of over 220 lbs on the face of the card! This is a far greater force than the mere 8 oz of force exerted by the water in the cup. However, since air is invisible, it appears that only the card is supporting the water in the cup. This demonstrates that the invisible air around us is capable of exerting a tremendous force.

Air Pressure – Experiment # 3:

BREAKING A YARDSTICK

Objective: To break a yardstick in half using air pressure.

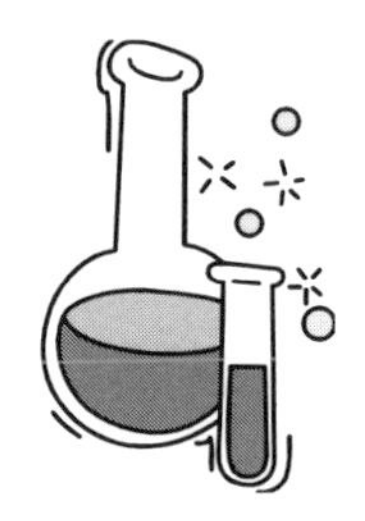

Materials:

- Cheap yardstick (available from a hardware store)
- Newspaper

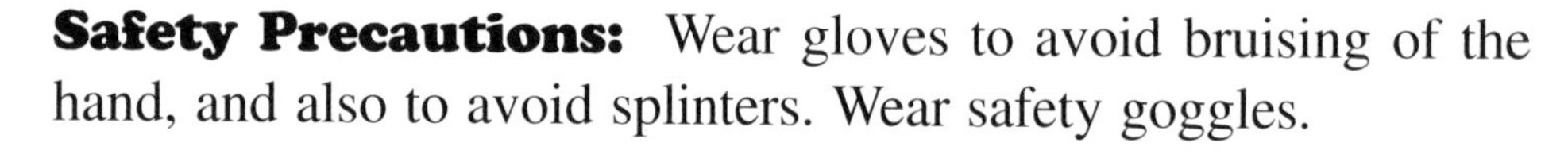

Safety Precautions: Wear gloves to avoid bruising of the hand, and also to avoid splinters. Wear safety goggles.

Procedure:

1. Place a yardstick on a table so that a 4-6 inch section of the yardstick is extending over the end of the table.
2. Place a large sheet of newspaper over the yardstick, arranged so that the fold in the paper is over the center of the yardstick. Smooth out the paper with your hand.
3. With a quick jerk of the hand, deliver a "karate chop" style blow to the yardstick. The key is to deliver the blow as close to the table as possible, without actually hitting the table.
4. The yardstick should break cleanly in half, with minimal movement of the newspaper. You may need to practice to perfect the proper technique.

Explanation: Air pressure is acting on the surface of the newspaper – with a force of over 9,000 lbs! Since the surface of the newspaper is approximately 621 square inches, and air pressure acts with a force of 14.7 lbs per square inch, an incredible amount of force is exerted on the surface of this newspaper. At over 9,000 lbs, this is roughly equivalent to that of two cars being parked on the yardstick! This huge force acting on the paper prevents the yardstick from moving when it is broken in half.

Air Pressure – Experiment # 4:

BALLOON IN A BOTTLE

Objective: To force a water-filled balloon into a bottle without touching it.

Materials:

- Balloon
- Matches
- Newspaper
- Empty wide-mouthed jar (a large empty pickle jar works well)
- Drinking straw

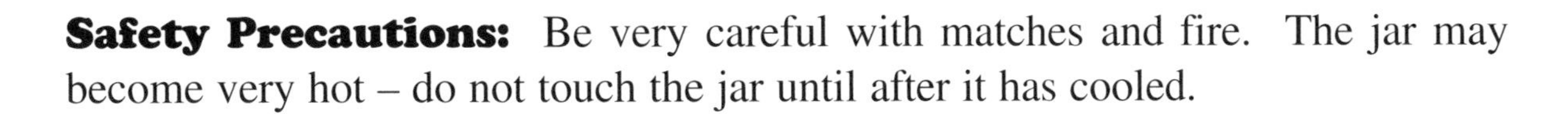

Safety Precautions: Be very careful with matches and fire. The jar may become very hot – do not touch the jar until after it has cooled.

Procedure:

1. Fill a balloon with water and tie it off, so that the diameter of the balloon is just slightly larger than the mouth of the bottle. The balloon should rest on the mouth of the bottle without falling in.
2. Remove the balloon, and place a loose wad of newspaper in the bottom of the jar.
3. Using a long match, set the newspaper on fire.
4. As soon as the flames begin to recede, place the balloon on the mouth of the jar. What happens to the balloon?

Explanation: The balloon will appear to be "sucked" into the jar, and will fall to the bottom. If this does not happen, you will need to make some adjustments. The balloon may be too large. The balloon may have been placed on the mouth of the jar either too late or too soon. Be patient and experiment – it will be well worth your time.

After the balloon is in the jar, it may be difficult to get it back out. Insert a drinking straw into the jar as you attempt to remove the balloon. The balloon will now lift out of the jar with relative ease.

As the air in the jar is heated, its molecules move faster and expand, thus causing much of the air to leave the jar. This creates a region of reduced pressure in the jar. There are now fewer air molecules in the jar than previously. As the fire goes out and this air in the jar now cools, air pressure within the jar is much less than outside the jar, since much of the air has escaped. Since the outside air pressure is greater, the balloon is pushed (not sucked!) into the jar.

It may be tempting to think that since fire consumes oxygen, then the consumption of the oxygen by the flames lowered the pressure in the jar. In actuality, as oxygen is consumed other gases, namely CO_2 and H_2O, are produced at the same rate. So this explanation is not adequate to explain the drop of pressure in the jar.

Related Experiments: A hard-boiled egg may be substituted for a water-filled balloon. This will necessitate a smaller-mouthed bottle.

Air Pressure – Experiment # 5:
VACUUM IN A CUP

Objective: To cause two cups to adhere to the sides of a balloon by reducing the air pressure within each cup.

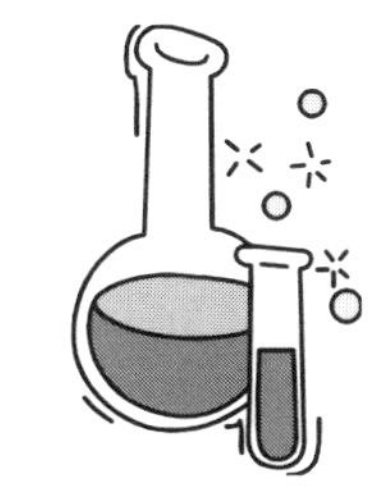

Materials:
- Balloon
- Two disposable plastic cups

Safety Precautions: None

Procedure:
1. Hold two cups together so their openings face each other.
2. Place a balloon between the two cups and blow up the balloon, without touching the cups.
3. Both cups should adhere to the balloon. It may take a little practice to position the cups properly.

Explanation: As the balloon is inflated, it pushes some of the air out of the cups, creating a region of reduced pressure within each cup. Since outside air pressure is now greater than the pressure of the air within each cup, the cups will adhere to the balloon without falling. They can actually be quite difficult to remove!

Air Pressure – Experiment # 6:

ADVENTURES WITH A DRINKING STRAW

Objective: To demonstrate that liquids cannot be "sucked" up with a straw.

Materials:

- Drinking straw
- Cup
- Scissors

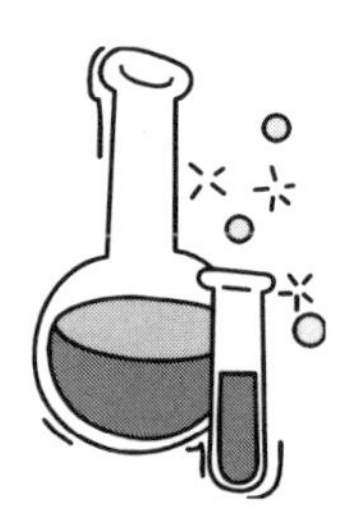

Safety Precautions: None

Procedure:

1. Put the straw into a cup of water and drink from the straw.
2. Now remove the straw. Using your scissors cut out a small hole in the straw, above the water line.
3. Replace the straw and attempt to drink. You will be unable to drink water with this "defective" straw!

Explanation: It is tempting to think that when you drink through a straw, you are sucking fluid up the straw and into your mouth. This is not what happens. There is no such thing as suction! As you expand your lungs, you increase their volume. This increase in volume causes the air within your lungs to exert less pressure. The air in the straw is now under greater pressure than the air in your lungs. As a result, air will travel from the straw to your lungs in an attempt to equalize the pressure. This creates a region of reduced pressure within the straw. As a result, the air pressure pushing down on the surface of the water is now greater than the air pressure within the straw. Therefore water is pushed up into the straw and into your mouth.

When a hole is made in the straw, air will flow into the straw, making the pressure inside the straw the same as that outside, rendering it useless.

Air Pressure – Experiment # 7:
CRUSHING A 2-LITER BOTTLE

Objective: To crush a 2-Liter bottle using only air pressure.

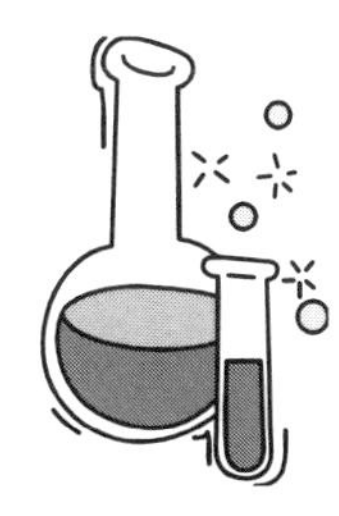

Materials:
- 2-Liter bottle

Safety Precautions: None

Procedure:
1. Remove the air from a 2-Liter bottle by "sucking" on it vigorously.
2. Observe what happens to the bottle.

Explanation: This very simple experiment illustrates a very profound principle. There is no such thing as suction. You are not sucking the bottle inwards. By removing some of the air from the bottle by expanding your lungs, a region of reduced pressure is created within the bottle. Since outside air pressure is greater than the air pressure inside the bottle, the bottle will be pushed inwards. If you can attach a vacuum cleaner hose to the bottle, the collapse of the bottle will be much more dramatic.

Air Pressure – Experiment # 8:

ALTITUDE AND AIR PRESSURE

Objective: To demonstrate that air pressure decreases with an increase in altitude.

Materials:

- 2-Liter bottle
- Nail
- Duct tape

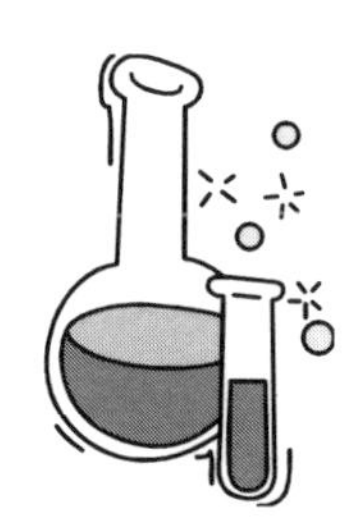

Safety Precautions: None

Procedure:

1. Using a nail, carefully poke 3 small holes in a vertical row along the side of a 2-Liter bottle. Poke one hole near the bottom, another in the middle, and the third near the top.
2. Cover each hole with a piece of duct tape.
3. Fill the bottle with water, and simultaneously remove all 3 pieces of duct tape. Observe.

Explanation: The water streaming out of the bottom hole should travel the greatest distance, and the water streaming from the top hole should travel the least distance. Air and water are both fluids, and thus share many of the same properties. The water travels the farthest from the bottom hole because it is under the most pressure due to the weight of the water above it.

Our atmosphere behaves the same way as the water within the bottle. Our atmosphere is essentially an ocean of air extending upward several hundred miles from the earth's surface. If you are standing at the bottom of this ocean of air (sea level), a much greater weight of air bears down on you than if you were to stand on top of a mountain. Since there is less air above you as your altitude increases, the pressure that this air exerts is also less. This explains why it is difficult to breath at very high altitudes, why water boils at lower temperatures on top of a mountain, and why your ears pop when you travel in an airplane.

Air Pressure – Experiment # 9:
"LEVITATING" A PERSON

Objective: To "levitate" a person using only air.

Materials:

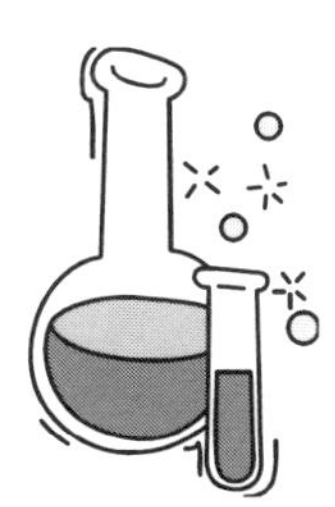

- Large trash bag
- 10 drinking straws
- Duct tape
- 13 people

Safety Precautions: Two responsible individuals should act as spotters and stand on either side of the person being "levitated."

Procedure:

1. Poke four of the straws an equal distance apart along the seam of one side of the trash bag. Allow approximately an inch or so of the straw to remain inside the bag.
2. Repeat the above procedure with the other side of the bag, and then insert two straws into the sealed (bottom) end of the trash bag.
3. Using duct tape, completely seal the open end of the bag. Also tape around each straw with the duct tape so as to make airtight seals.
4. Place the bag on a table.
5. Have a volunteer sit on the bag and bring his knees toward his chest, so that his entire body is on the bag.
6. Have 10 people blow into the straws until the bag is completely inflated and the volunteer rises into the air.
7. If the bag tears, simply repair with duct tape – this will serve to make the bag stronger in the future.

Explanation: As the bag inflates with air, it begins to exert its incredible power. The same method has been used to free the wreckage from atop the victims of automobile crashes. A flat, deflated, heavy duty "balloon" is inserted underneath the wreckage and then pumped up with air. The bag you inflated is capable of lifting over 400 lbs!

Air Pressure – Experiment # 10:
DOES AIR TAKE UP SPACE?

Objective: To demonstrate that air has volume and takes up space.

Materials:
- Tissue
- Drinking glass
- Pan

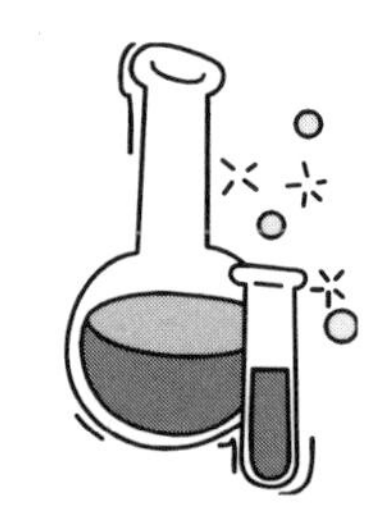

Safety Precautions: None

Procedure:
1. Wad up a piece of tissue and place it at the bottom of a glass.
2. Invert the glass straight down into a pan of water until it touches the bottom.
3. Remove the glass and then remove the tissue. Observe the tissue.

Explanation: When the tissue is removed, it will be completely dry. Since the glass is filled with air, water cannot enter the glass to make the tissue wet. This demonstrates that air has volume and takes up space. Air and water cannot occupy the same space. If you were to tilt the glass so that air could escape, then water would enter and wet the tissue.

Air Pressure – Experiment # 11:
BLOWING UP A BALLOON IN A BOTTLE

Objective: To discover the power of the air.

Materials:

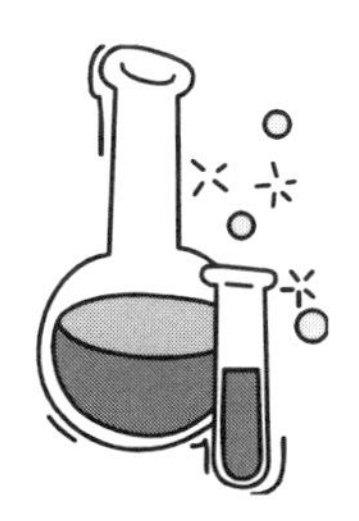

- Balloon
- 2-Liter bottle
- Straw

Safety Precautions: None

Procedure:

1. Place a balloon completely inside of a 2-L bottle, except for the lip of the balloon, which you will need to blow into.
2. Attempt to blow up the balloon. Are you successful?
3. Now insert a straw into the mouth of the bottle as you attempt to blow up the balloon. What happens?

Explanation: The bottle is not empty – it is full of air. This air is capable of exerting a tremendous force – a much greater force than your lungs are capable of exerting. When you attempt to blow up the balloon, the balloon pushes up against the air in the bottle. The air then pushes back. This makes it very difficult to blow up the balloon inside the bottle.

By inserting a straw into the bottle, air can leave the bottle as the balloon inflates. Therefore the balloon can easily be blown up within the bottle. The same effect can be accomplished by drilling a hole in the bottle, which allows air to escape as the balloon is inflated.

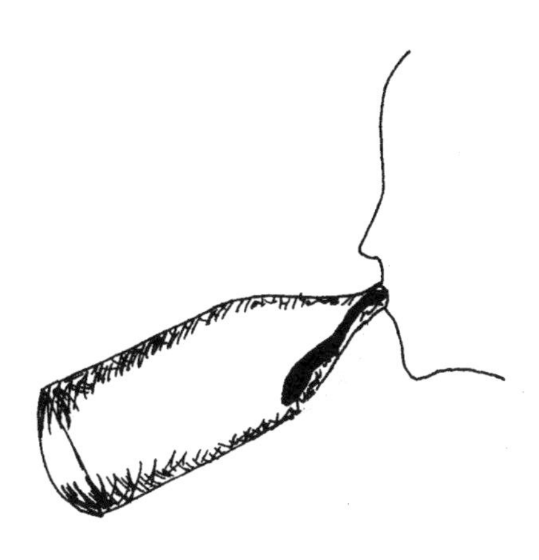

Air Pressure – Experiment # 12:

MAGIC SODA CANS

Objective: To demonstrate Bernoulli's Principle.

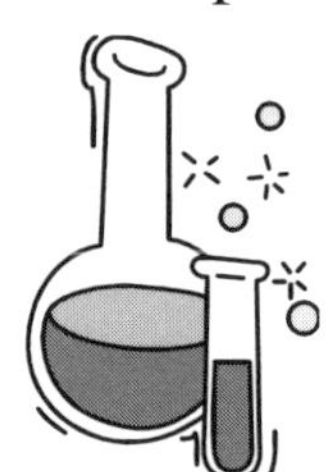

Materials:

- Two empty soda cans

Safety Precautions: None

Procedure:

1. Place two soda cans on their sides on a table approximately one inch apart.
2. Blow straight downward between the two cans. What happens?

Explanation: The cans should move together as you blow between them. This is verification of Bernoulli's Principle, which states that moving air exerts less pressure than still air. Since we are blowing in the middle of the two cans, we are creating a region of lower pressure. Outside air pressure acting on both sides of the cans is therefore greater than the pressure in the middle of the cans. This greater outside air pressure serves to push the two cans together.

Bernoulli's Principle is named after Daniel Bernoulli, a Swiss scientist of the 18th century. He made many important discoveries concerning the behavior of fluids (liquids or gases) under pressure.

Air Pressure – Experiment # 13:
THE AMAZING PLUNGER

Objective: To demonstrate the power of air pressure using a toilet plunger.

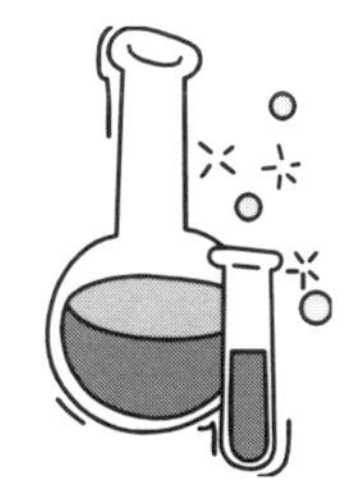

Materials:

- Unused toilet plunger

Safety Precautions: None

Procedure: Use the plunger to lift up a variety of objects: plates, boards, books, etc.

Explanation: There is perhaps no better instrument to illustrate the power of air pressure than an ordinary plunger, which is really just a very large "suction" cup. But remember, there is no such thing as suction. The plunger works by pushing away some of the air from underneath it, thereby creating a region of reduced pressure between itself and the object. Air pressure acting on the top of the plunger serves to create a fairly tight seal that enables the plunger to lift objects. Remember that nothing is sucking; the air is simply pushing downward on the plunger.

If you have ever used a plunger to unclog a toilet, you may have noticed bubbles of air coming up from below the water as the plunger is being used. By reducing the air pressure above the water, the air pressure in the pipe below the water is now greater. It is this greater air pressure pushing upward that actually unclogs the toilet.

Air Pressure – Experiment # 14:

HOW TO GET A BALLOON IN A BOTTLE

Objective: To force a balloon inside of a bottle using air pressure.

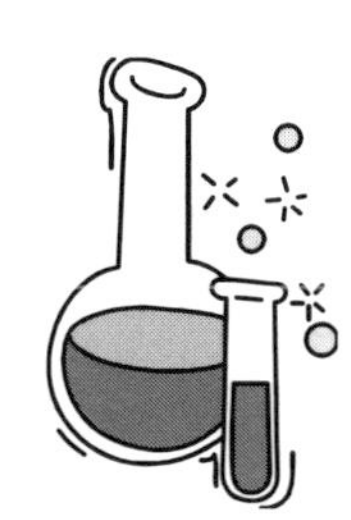

Materials:

- 32 oz lemon juice bottle (or equivalent)
- Balloon
- Teakettle
- Stove or hot plate

Safety Precautions: Exercise caution when using the stove and when handling boiling water. Make sure you use a sturdy glass bottle that can withstand the boiling water. Place the bottle in the sink when pouring the boiling water. Use potholders to handle the bottle of hot water. Wear safety goggles.

Procedure:

1. Heat 1 Liter of water to boiling.
2. Fill the bottle with this boiling water, then immediately empty it.
3. Immediately place a balloon over the mouth of the bottle. Observe.

Explanation: This experiment is well worth taking the time to master, since the results are so dramatic. By filling the bottle with boiling water and emptying it, the air in the bottle is heated. This causes the air to expand, with much of it leaving the bottle. When the air in the bottle cools, it exerts less pressure than outside air pressure, since much of the air previously left the bottle. Since outside air pressure is now greater, it pushes down on the balloon, causing it to completely inflate inside of the bottle. Show your friends the balloon in the bottle – without telling them how you did it – and then challenge them to duplicate this feat.

Air Pressure – Experiment # 15:

CREATING A SIPHON

Objective: To demonstrate how a siphon works.

Materials:

- Two 2-Liter bottles
- Three feet of plastic aquarium tubing
- Food coloring

Safety Precautions: None

Procedure:

1. Fill one bottle with water, add a few drops of food coloring, and place on a table.
2. Place the other bottle on the floor.
3. Insert the tubing in the bottle of water and begin removing the air from the tubing with your mouth.
4. When enough air is removed from the tubing, water will begin to come out of the tubing. At this point, quickly insert the end of the tubing into the empty bottle.
5. As long as the empty bottle stays below the level of the filled bottle, the siphon will continue to operate, and one bottle will completely empty into another. If the empty bottle is raised above the level of the first bottle, the siphon will cease to work.

Explanation: As the air is removed from the tubing, a vacuum is created in the tubing, much the same way as drinking from a straw. Therefore, air pressure pushing on the surface of the water forces the water up the tube and into the other bottle. Gravity will keep the siphon operating, but this siphon cannot operate against gravity and cause water to flow uphill if the position of the two bottles is switched.

Air Pressure – Experiment # 16:
THE MYSTERIOUS PING-PONG BALL

Objective: To cause a ping-pong ball to descend to the bottom of an aquarium without touching it, and to cause the same ping-pong ball to rise up out of the aquarium without touching it.

Materials:
- Two gallon aquarium or larger
- 2-Liter bottle
- Scissors
- Ping-pong ball

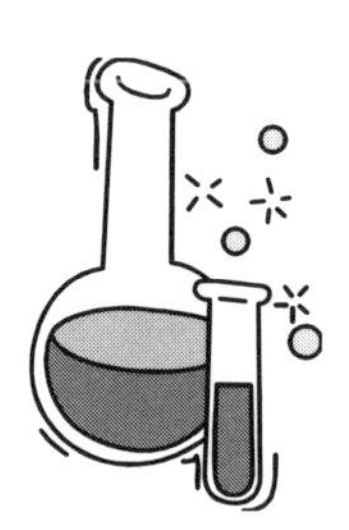

Safety Precautions: None

Procedure:
1. Using scissors, cut the bottom from the 2-Liter bottle.
2. Fill an aquarium about ¾ full with water.
3. Place the ping-pong ball on the surface of the water. With the lid on the bottomless 2-Liter bottle, place it over the ping-pong ball and push downward. The ball will be pushed to the bottom.
4. Now take the cap off, and allow water to fill in the bottle. When the water and the ball have reached the top of the bottle, replace the cap, and pull the bottle out of the water. As long as the bottle does not come completely out of the water, the ping-pong ball will rest on top of the water at a height above the surface of the water within the aquarium!

Explanation: Air takes up space. Since the bottle is filled with air, water cannot enter the bottle, and this column of air forces the ping-pong ball to the bottom of the aquarium. When the ball rises up out of the water, it is floating on top of the water in the bottle. The water does not fall out of the bottle due to the buoyant force of the water below it. Water acts like the air in that it exerts pressure in all directions, even upward.

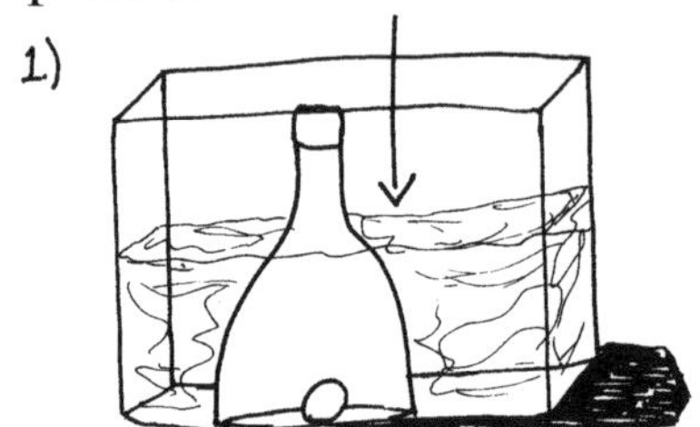

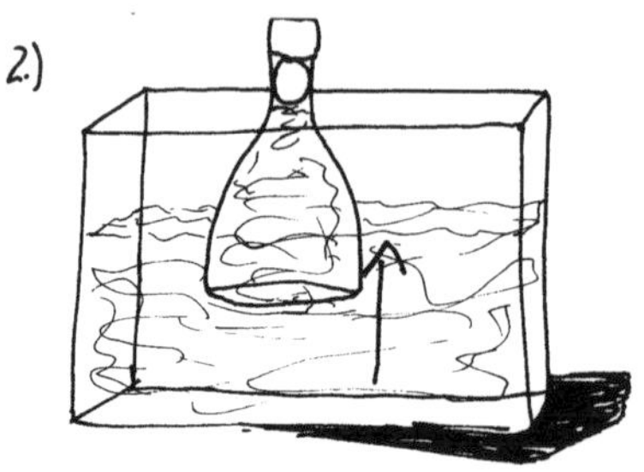

Air Pressure – Experiment # 17:

THE MYSTERIOUS BALLOONS

Objective: To discover how air can flow from an area of low pressure to one of high pressure.

Materials:

- Two balloons
- Empty thread spool

Safety Precautions: None

Procedure:

1. Blow up one balloon to about ¾ its full size, and carefully attach to one end of the spool. Pinch the balloon to prevent it from losing air.
2. Blow up the other balloon to about half the size of the first balloon, and attach to the other end. Pinch it to prevent air from escaping.
3. Predict what will happen to the two balloons if you release your fingers and allow air to flow freely between them.
4. Release your fingers and observe. After a short time, you will notice that the large balloon gets larger, and the small balloon gets smaller!

Explanation: This is an excellent example of a discrepant event. A discrepant event occurs when you are expecting one thing to happen but the opposite occurs. In this experiment, it is tempting to propose a more complex explanation than what actually occurs.

The smaller balloon, since it is not inflated as much, has greater elasticity, and thus is able to exert a greater force on the air within it. The larger balloon, since it is inflated so much, is stretched out, and therefore cannot exert a very strong force on the air within it. Since the rubber of the smaller balloon is more elastic, it is able to force its air into the larger balloon.

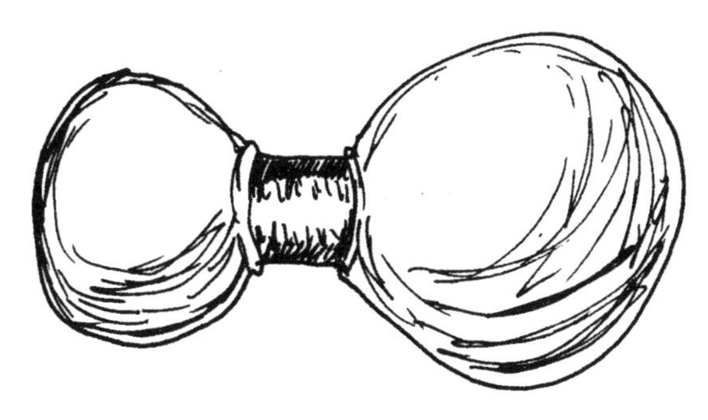

Air Pressure – Experiment # 18:
FORCING A STRAW THROUGH A POTATO

Objective: To force a straw through a potato using air pressure.

Materials:
- Straw
- Potato

Safety Precautions: None

Procedure:
1. Try to force a straw through a potato, making sure to leave the end of the straw uncovered. What happens?
2. Now, holding your thumb over the end of the straw, attempt to force the straw through the potato. Now what happens?

Explanation: The straw is much too flimsy to penetrate the potato by itself. However, by holding your thumb over one end of the straw, the straw is now filled with air. This column of air represents a formidable force that allows the straw to easily pass through the potato. It is almost the same as penetrating the potato with a solid rod, which would not be too difficult. This demonstrates the extraordinary power of the invisible air that surrounds us.

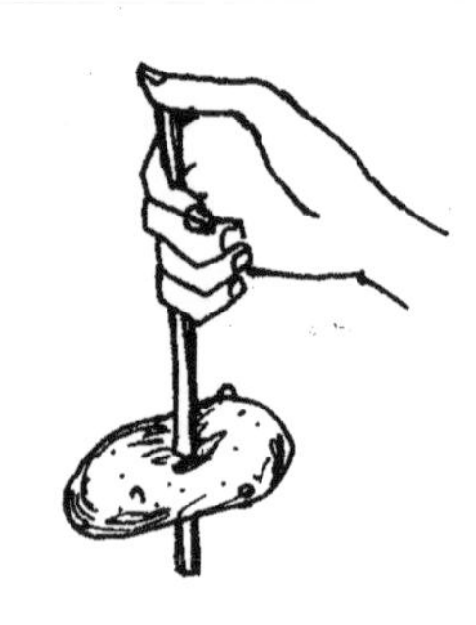

CHAPTER 3
SURFACE TENSION

Surface tension enables us to skip stones across a lake and fill a water glass above the brim. It is also responsible for the shape of raindrops and the condition of our hair when we get out of the swimming pool. Surface tension acts like an "invisible" skin on the surface of liquids, allowing insects such as water striders to walk on water.

Surface tension is due to the difference between the attractions among water molecules at the surface and those within the body of a liquid. Within a liquid, water molecules are pulled evenly in all directions by neighboring water molecules. At the surface, however, water molecules only experience a downward pull. As a result, all of the water molecules become bunched together at the surface, creating an invisible "skin."

Water has a very high surface tension because it is polar. Polar molecules exert strong intermolecular attractions on other water molecules. The polarity of water causes it to be very sticky, enabling water molecules to stick very well to one another, and also to many other substances.

In this chapter, we will explore the fascinating world of surface tension . . .

Surface Tension – Experiment # 1:
THE MAGIC BOTTLE

Objective: To demonstrate that the cohesiveness of water droplets can lead to surface tension.

Materials:
- 1-Liter plastic bottle (any size will work – a firm type of plastic works best)
- Piece of fine mesh window screen
- Rubber band
- Nail
- 3 x 5 inch index card

Safety Precautions: None

Procedure:
1. With a nail, punch a hole in the side of the bottle near the bottom.
2. Place the screen tightly across the mouth of the bottle, fastening it down with a tightly drawn rubber band.
3. Fill the bottle all the way with water.
4. Put the card over the mouth of the bottle.
5. Put your finger over the hole.
6. With your hand on top of the card, carefully turn the bottle upside-down.
7. With your finger still on the hole, remove your hand and the card. The water should stay in the bottle.
8. Now remove your finger from the hole, and the water will flow out from the top of the bottle. Put your finger back over the hole, and the water will stop.

Explanation: Water molecules tend to stick together very well. This is due to the fact that water is a polar molecule, meaning that it has both a positive and a negative end. Since opposites attract, the positive end of one water molecule sticks to the negative end of another molecule. This is why water forms such large droplets. Two like molecules sticking together is known as cohesion.

In the above experiment, the water droplets fill in the spaces between the holes in the screen, due to the cohesiveness of the water molecules. The force that all of the water molecules collectively exert is known as surface tension.

The force of surface tension is strong enough to hold up the water in the bottle. However, when your finger is removed from the hole, air enters the bottle, and the force of air pushing down on the water overcomes the surface tension, thus allowing the water to flow out of the bottle.

Surface Tension – Experiment # 2:

ADVENTURES WITH PEPPER

Objective: To demonstrate the effect of the surface tension of water on pepper, and then discover a way to destroy this surface tension.

Materials:

- Cup
- Pepper
- Dishwashing liquid
- Toothpick

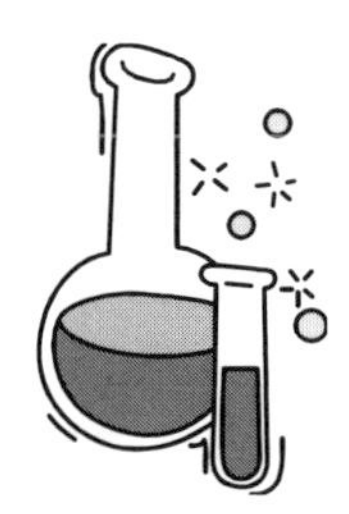

Safety Precautions: None

Procedure:

1. Fill a cup with tap water.
2. Sprinkle some pepper on the surface of the water. The pepper should float.
3. Now, dip the toothpick in the dishwashing soap, and carefully touch it to the center of the surface of the water. Observe what happens.

Explanation: Pepper is denser than water and should sink. However, due to the surface tension of water, it floats. When the soap is placed in the water, the soap molecules immediately bond to the water molecules, destroying the cohesiveness of the water and thus its surface tension. The soap causes the particles of pepper to immediately disperse from the surface of the water to the edges. The pepper particles that disperse are following the path of the soap molecules as they are being dispersed in the water. Eventually, the soap will disperse across the entire surface of the water, causing most of the pepper particles to sink.

Surface Tension – Experiment # 3:

A FLOATING PAPER CLIP

Objective: To demonstrate surface tension by floating a paper clip in water.

Materials:

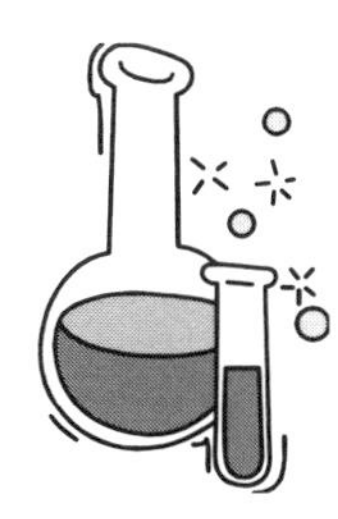

- Two paper clips
- Cup
- Dishwashing liquid

Procedure:

1. Bend the inner loop of a paper clip so it forms a 90° angle with the outer loop. It can now be used as a paper clip holder.
2. Fill a cup with water.
3. Place the paper clip holder underneath the other paper clip, then gently place it into the cup of water until it floats on the surface. Remove the paper clip holder. Be patient – with a little practice this can be easily done.
4. When you tire of watching the floating paper clip, add a drop of dishwashing liquid anywhere on the surface. The paper clip will immediately sink.

Explanation: Even though the paper clip is denser than water and should sink, the surface tension of the water causes the paper clip to float. The dishwashing liquid binds with the water molecules, destroying the cohesiveness of the water, thus destroying its surface tension. The paper clip then sinks.

Surface Tension – Experiment # 4:

DROPS ON A PENNY

Objective: To demonstrate the concept of surface tension by placing drops of water on a penny.

Materials:
- Penny
- Eyedropper

Safety Precautions: None

Procedure:
1. Predict how many drops of water can be placed on a penny before the water flows off.
2. Using an eyedropper, carefully place drops of water on a penny, until the water flows off.
3. Compare your results with your prediction.
4. Have a contest to see who can balance the most drops on a penny. To reduce the variables, have each person hold their dropper completely perpendicular to the penny and also the same distance from the penny. Also make sure each penny is facing the same way (heads or tails).

Explanation: Water is very cohesive – its molecules stick to each other very well. This leads to surface tension, which is why such an incredible amount of water can be balanced on a penny. With practice you can get over 50 drops to adhere to a single penny!

Surface Tension – Experiment # 5:
HOW MANY PAPER CLIPS CAN FIT INTO A FULL CUP OF WATER?

Objective: To demonstrate surface tension by determining the number of paper clips that can be placed in a cup of water before it overflows.

Materials:
- Cup
- Paper clips

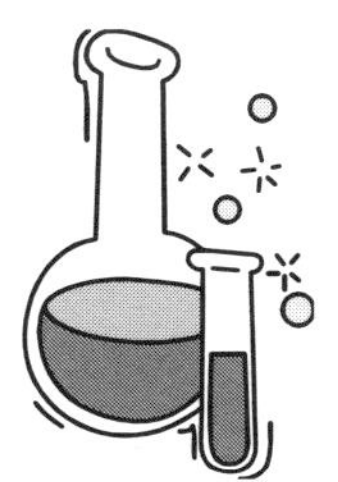

Safety Precautions: None

Procedure:
1. Fill a cup to the brim with water.
2. Predict the number of paper clips that you will be able to drop into the cup of water before it overflows.
3. Carefully drop in paper clips, one at a time, until the water overflows.
4. Compare your prediction with the actual number of paper clips needed.

Explanation: You will probably be amazed at the number of paper clips that you can fit into a "full" glass of water. What happens to the water that is displaced by the paper clips in the cup? Since water molecules are attracted to one another, the water level will actually end up being above the brim of the cup. Each paper clip displaces a small amount of water, and due to the surface tension of the water, the water rises up and out of the cup before overflowing.

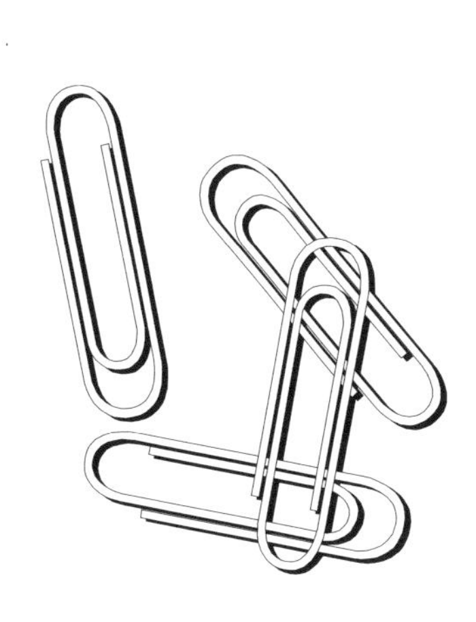

CHAPTER 4
GASES, LIQUIDS, and SOLIDS

Gases represent a state of matter in which the molecules are moving very rapidly, thus they always expand to fill their container. They exert pressure by constantly colliding with the walls of their container. Gases are much less dense than either liquids or solids, and are usually invisible.

Liquids are unique in that they share some common properties that are quite different from gases and solids. Liquids have a definite volume, but not a definite shape. They do not expand to fill their containers like gases do, but they do assume the shape of their container.

Solids have a definite shape and a definite volume. They cannot readily be compressed. Solids tend to be more dense than liquids or gases, although there are exceptions. All metals are solid at room temperature, except for mercury. Solids tend to have fairly strong bonds between their atoms, which make them rigid.

In this chapter, we will study all three phases of matter, as well as changes that occur between phases . . .

Gases, Liquids, and Solids – Experiment # 1:

HOW TEMPERATURE AFFECTS THE SPEED OF MOLECULES

Objective: To determine how temperature affects the speed of molecules.

Materials:

- Two heat-resistant glass containers
- Stove
- Food coloring
- Ice cubes

Safety Precautions: Exercise caution when heating water and when handling hot water.

Procedure:

1. Heat up approximately 200 mL of water to just below boiling. Pour into the first container.
2. Into a second container, pour some cold water from the tap, and add ice cubes. Remove the ice cubes after approximately one minute.
3. Add a single drop of food coloring to each glass. Observe.

Explanation: This experiment vividly demonstrates the effect of temperature on the speed of molecules. Temperature measures how fast the molecules in a substance are moving. The food coloring diffuses much faster in the hot water because their molecules are moving much faster than the molecules of cold water. In the cold water the molecules are moving slower, therefore the food coloring takes much longer to diffuse.

Gases, Liquids, and Solids – Experiment # 2:

OBSERVING PHASE CHANGES

Objective: To observe the temperature at which a solid turns into a liquid, and the temperature at which a liquid turns into a gas.

Materials:

- Ice cubes
- Pan
- Stove or hot plate
- Thermometer

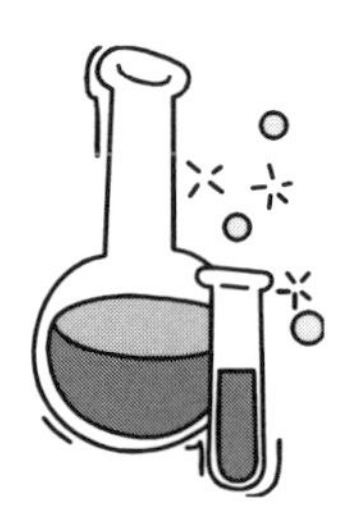

Safety Precautions: Exercise caution when using the stove, and also when handling boiling water.

Procedure:

1. Fill a small pan approximately half-full with ice cubes. Record the temperature of the ice.
2. Heat the pan of ice cubes gently until all of the ice melts. Leave the thermometer in the pan.
3. Record the temperature when the very last bit of ice melts.
4. Continue heating. Record the temperature when the water achieves a full boil. Discontinue heating after boiling is achieved.

Explanation: Melting, the act of changing from a solid to a liquid, should occur at approximately 0°C or 32°F. Your results may differ depending upon the purity of the water and the accuracy of your thermometer. Boiling, the act of changing from a liquid to a gas, should occur at approximately 100°C or 212°F, again depending on the purity of the water and the accuracy of your thermometer. Pure water under normal atmospheric pressure (1 atm) will have a melting point of exactly 0°C and a boiling point of exactly 100°C. This is what the Celsius temperature scale is based upon.

Gases, Liquids, and Solids – Experiment # 3:

HOW DOES SALT AFFECT THE FREEZING POINT OF WATER?

Objective: To observe how salt affects the freezing point of water.

Materials:

- Gallon freezer bag
- Quart freezer bag
- Crushed ice
- Salt
- Thermometer
- Gloves

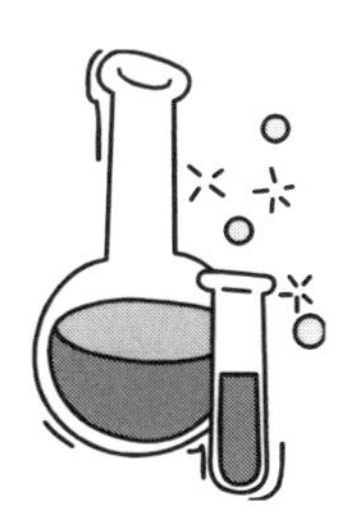

Safety Precautions: None

Procedure:

1. Fill the gallon freezer bag about halfway with crushed ice.
2. Add approximately one cup of salt. More can be added later if necessary.
3. Put on a pair of gloves and knead the ice and salt mixture through the bag until the ice has completely melted.
4. Record the temperature of the saltwater mixture. It should be below 0°C.
5. Place approximately one ounce or about 30 mL of water in the quart freezer bag. Seal this bag and place it in the larger freezer bag as soon as the ice has completely melted.
6. Allow the baggie of water to remain in the saltwater mixture until it turns into a solid. The water in the baggie will actually freeze into ice, surrounded only by salt water!

Explanation: The initial temperature of the ice is below 0°C, since ice melts at 0°C. As salt is added to the ice, it breaks apart the bonds that form the ice crystals, causing the ice to melt. Since melting is an endothermic process, energy is absorbed from the surroundings. As a result, the temperature of the melted ice will fall below 0°C. Since pure water freezes at 0°C, the baggie of water will freeze if placed in a solution which is below its freezing point.

The addition of salt to water lowers its freezing point. The particles of salt impede the formation of the bonds necessary for ice formation to occur, interfering with the crystallization process. This is one reason why the oceans seldom freeze. Actually, any substance that is dissolved in water will lower the freezing point of water. This lowering of the freezing point is known as freezing point depression.

Gases, Liquids, and Solids – Experiment # 4:

HOW DOES SALT AFFECT THE BOILING POINT OF WATER?

Objective: To observe how salt affects the boiling point of water.

Materials:

- Pan
- Stove or hotplate
- Thermometer
- Salt

Safety Precautions: Exercise caution when using the stove and when working with boiling water.

Procedure:

1. Place some water in a pan, and heat to boiling.
2. When the water is at a full boil, record its temperature.
3. Add a copious amount of salt and observe its immediate effect.
4. When the water again reaches a full boil, record the temperature.

Explanation: Pure water at sea level boils at 100°C. The addition of salt, or any other substance dissolved in water, will raise its boiling point. This is known as boiling point elevation. Water molecules are bonded to one another, which is why they must be heated in order for them to break free from one another and escape into the gaseous state. When dissolved in water, the salt particles also bond to the water molecules. Therefore, it requires more energy to break these bonds so that the water molecules can escape. In addition to this, the salt particles impede the escape of the water molecules, blocking their path as they attempt to escape. So in order to boil salt water, a higher temperature must be reached.

Gases, Liquids, and Solids – Experiment # 5:
DOES WATER EXPAND WHEN IT FREEZES?

Objective: To discover the effect of freezing on the volume of water.

Materials:
- 20 oz plastic soda bottle
- Freezer
- Permanent marker

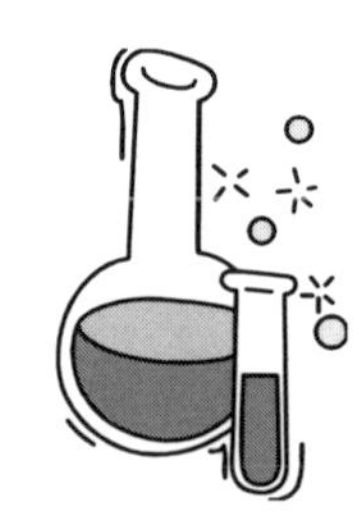

Safety Precautions: None

Procedure:
1. Fill the bottle about halfway with water and put the lid on tightly.
2. With a permanent marker, carefully draw a line representing the water level.
3. Place in the freezer overnight. Observe. Is the ice above the level of the line?

Explanation: Water is one of the few liquids that expands when it freezes. This is due to the unique crystal structure of ice. When water solidifies, the individual water molecules always arrange themselves into a hexagonal pattern, with a great deal of empty space in the middle. This causes ice to expand as it forms, making it less dense than liquid water. This explains why ice floats.

Gases, Liquids, and Solids – Experiment # 6:
POP YOUR TOP!

Objective: To observe the effect of the increase of gas pressure within an enclosed container.

Materials:

- Translucent or transparent film canister with snap-on lid
- Alka-Seltzer tablet

Safety Precautions: Do not look down upon the film canister after it is sealed. Do not point at anyone! Wear safety goggles.

Procedure:

1. Fill the film canister most of the way with water.
2. Drop in an Alka-Seltzer tablet.
3. Quickly put the lid back on. Stand back! The lid should pop up into the air several feet. If it does not, you may need to try a different film canister. The lid must snap on tightly.
4. Repeat the experiment, except this time invert the canister so that the lid is pointing down.

Explanation: When the Alka-Seltzer tablet is dropped into water, carbon dioxide (CO_2) gas is immediately released. You will hear it fizz. Since this gas cannot escape, pressure gradually builds up in the film canister until the lid is popped off.

When placed upside-down, the canister is shot up into the air. This demonstrates Newton's Third Law of Motion, which states that for every action there is an equal and opposite reaction. The downward force exerted by the escaping gas causes the canister to be projected upward. Rockets are launched using this same principle. The force of the gases being expelled upon ignition launches the rocket upward.

Gases, Liquids, and Solids – Experiment # 7:
A HOT AIR BALLOON

Objective: To observe the effect of heating a gas.

Materials:

- 2-Liter bottle
- Balloon
- Stove or hot plate
- Two pans
- Ice cubes

Safety Precautions: Exercise caution when using the stove and when boiling water.

Procedure:

1. Place approximately one inch of water in a pan and place on the stove.
2. Secure a balloon over the mouth of the 2-Liter bottle.
3. Place the balloon in the pan of water and heat until the water boils. The balloon should expand.
4. In another pan, place several inches of water and some ice cubes. Immediately transfer the bottle to this pan. The balloon should deflate.

Explanation: This experiment is an excellent demonstration of Charles's Law, which states that as the temperature of a gas increases, its volume increases. Conversely, as the temperature of a gas decreases, its volume decreases.

As the air in the bottle is heated, the molecules of air in the bottle move faster and expand, which inflates the balloon. As the air cools, the molecules slow down, causing the air in the bottle to contract, which causes the balloon to deflate.

Gases, Liquids, and Solids – Experiment # 8:

SLICING A BLOCK OF ICE

Objective: To observe the effect of pressure on the melting of ice.

Materials:

- Ice cube or large block of ice
- Thin copper wire
- Two one-pound weights
- Board

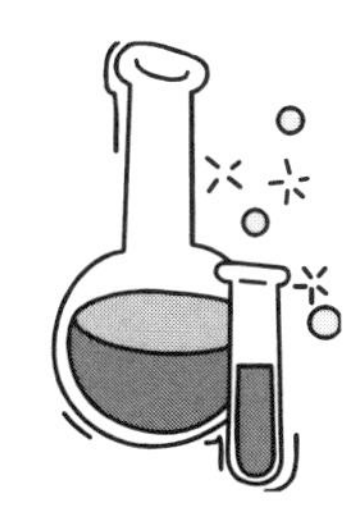

Safety Precautions: None

Procedure:

1. Place a board over your kitchen sink.
2. Tie an equal weight to each side of a one-foot long piece of thin copper wire.
3. Place the block of ice in the center of the board and place the wire with weights attached directly across the ice, so that it stays balanced.
4. Observe for a period of time until the wire has completely cut through the ice. The block of ice will refreeze as the wire passes through it.

Explanation: As pressure is put on the ice, the ice melts. This water from the melted ice is surrounded by the ice cube, which immediately refreezes the water. This is why the wire passes completely through the ice without slicing the block of ice in half.

This is an excellent demonstration of LeChatelier's Principle, which states that if pressure is applied to any system, that system will do whatever it can to relieve that pressure. For example, if pressure is applied to an inflated balloon by poking it with your finger, the balloon relieves that stress by deforming inward, or reducing its volume. In the same way, if stress is applied to ice, the best way to relieve that stress is to reduce its volume. As water freezes into ice it expands or increases its volume, so as ice turns into water its volume decreases. This decrease in volume is a way for the ice to relieve the pressure that is being applied to it. This is what happens when skaters skate on ice. The pressure of the skate immediately melts the ice, and the skater is actually skating on a thin film of water. Ice that is too cold is harder to melt, and thus more difficult to skate upon. Since this melted water is in direct contact with the ice below it, it immediately refreezes.

Gases, Liquids, and Solids – Experiment # 9:
HOT AND COLD AIR FROM THE SAME MOUTH

Objective: To demonstrate that gases can change temperature depending on the size of the opening through which they are released.

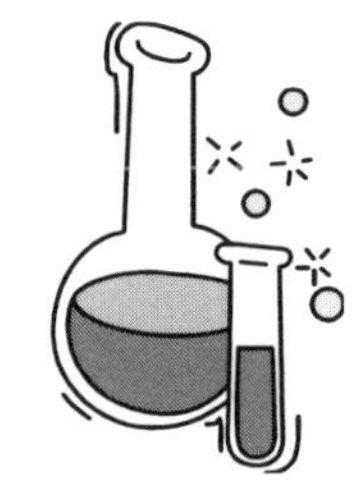

Materials: None

Safety Precautions: None

Procedure:

1. Purse your lips to make a narrow opening, and then blow into your hand. It should feel cool.
2. Open your mouth wide and then blow into your hand. It should feel hot.

Explanation: How can cold air and hot air come from the same mouth, where the temperature is approximately 98.6°F? The size of the opening is the key. As gas is forced through a small opening, its speed increases. This causes moisture on the surface of your skin to evaporate more quickly. Since evaporation cools your skin, this fast moving air feels cooler. When your mouth is open wide, the air is moving much more slowly so the air feels hot as it comes from your mouth. Since little evaporation occurs, you do not experience cooling on your skin.

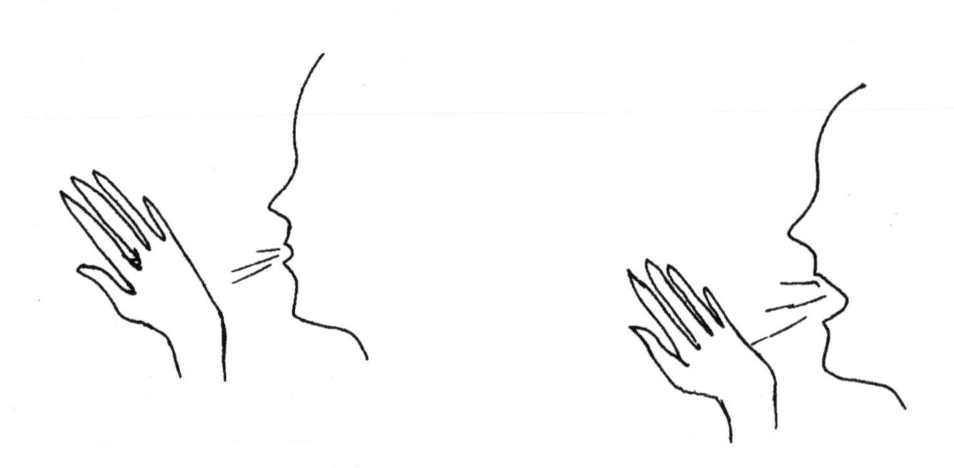

Gases, Liquids, and Solids – Experiment # 10:
BOYLE'S LAW IN A BOTTLE

Objective: To demonstrate Boyle's Law by inflating a balloon inside of a bottle.

Materials:

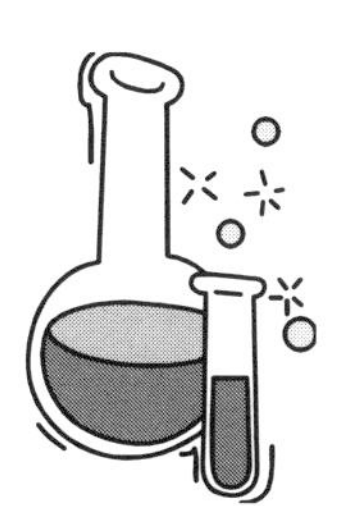

- Glass pickle jar
- Balloon
- Drinking straw
- Modeling clay
- Drill

Safety Precautions: Exercise caution when using drill.

Procedure:

1. Blow up a balloon several times and deflate it, so as to stretch out the rubber as much as possible.
2. Inflate the balloon to about a fourth of its maximum volume, and tie it off. Place it within the jar.
3. Drill a hole in the lid of the jar so that a drinking straw just fits through it. The size of the hole will vary, depending upon the diameter of the straw.
4. Place the lid on the jar, insert the straw through the hole, and seal the hole with modeling clay.
5. Remove some air from the jar by inhaling on the straw. If you need to take a breath, pinch on the straw, then resume.
6. You should notice that the balloon will inflate. If not, you will need to check that the seal is tight around the straw. A smaller bottle means less air needs to be removed, so better results can be obtained.
7. After the balloon inflates, let air back into the bottle. You will then notice the balloon deflate.

Explanation: Boyle's Law states that as the pressure on a gas increases, its volume will decrease. Therefore, as the pressure on a gas decreases, its volume increases. Volume and pressure are inversely proportional, assuming constant temperature. As the air is removed from the jar, the pressure around the balloon decreases, therefore the volume of the balloon increases. As air is let back into the jar, the pressure around the balloon increases, therefore its volume decreases and the balloon assumes its original size.

It helps to understand this experiment if we understand the nature of gas molecules. The balloon is filled with molecules of oxygen and nitrogen from the air, each colliding against the inner walls of the balloon many times each second. This is what causes the balloon to expand as you initially inflate it with your lungs. However, the balloon can only expand to a certain volume, because air pressure is pushing on the outside of the balloon from all directions. As the air is removed from the bottle, air pressure within the bottle decreases, so there are less molecules of air pushing in on the balloon from the outside. Therefore, the gas molecules inside the balloon, which are vigorously colliding against the inside walls of the balloon, can inflate the balloon even more. When air is introduced back into the jar, the additional outside air molecules pushing in on the balloon force it to assume its original volume.

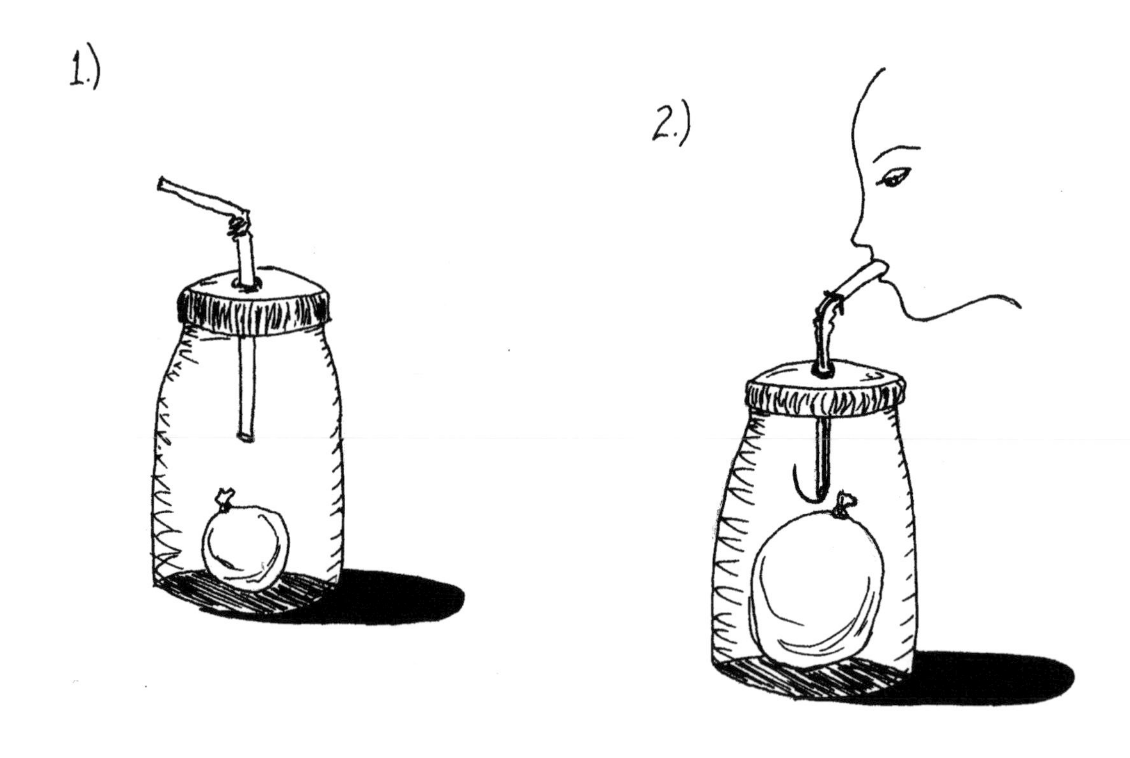

Gases, Liquids, and Solids – Experiment # 11:
BOILING WATER IN A SYRINGE

Objective: To boil water at a temperature much lower than 100°C.

Materials:

- Stove or hot plate
- Syringe with cap
- Pan
- Thermometer

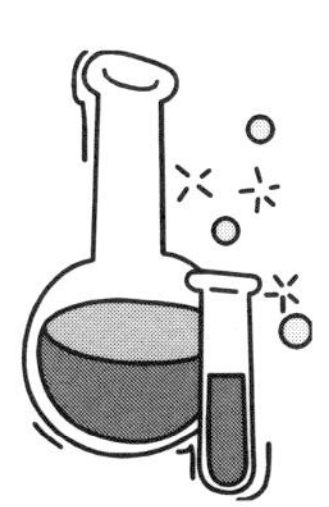

Safety Precautions: Exercise caution when using stove and when handling hot water.

Procedure:

1. Heat a cup of water to approximately 70°C.
2. Remove the cap from the syringe, and completely depress the plunger so as to push all of the air from the syringe.
3. Place the syringe in the hot water, and lift up on the plunger so the syringe fills about halfway with hot water.
4. Replace the cap on the syringe, and lift up on the plunger so as to produce as much empty space as possible above the water.
5. You should notice that as the plunger is lifted up, the water immediately begins to boil, at a temperature far below the normal boiling point of water.

Explanation: Molecules of water are always trying to escape from the liquid state into the vapor state. This is why a glass of water will evaporate if left uncovered. When water boils, many water molecules are escaping quickly. The pressure of the air above prevents water molecules from escaping more quickly. As water is heated, the water molecules gain enough energy to break free from the liquid and escape as a vapor.

However, by reducing the number of air molecules above a liquid, its boiling point will be lowered, because it will now be easier for the water molecules to escape. That is why water boils at a lower temperature at higher elevations.

When the plunger of the syringe is lifted above the water, a near vacuum is created above the water. Since there are fewer air molecules now above the water's surface, the water molecules find it much easier to escape. Therefore the water boils.

Gases, Liquids, and Solids – Experiment # 12:
BALLOON IN A FREEZER

Objective: To demonstrate Charles's Law by placing a balloon in a freezer.

Materials:
- Balloons
- Freezer
- String
- Ruler

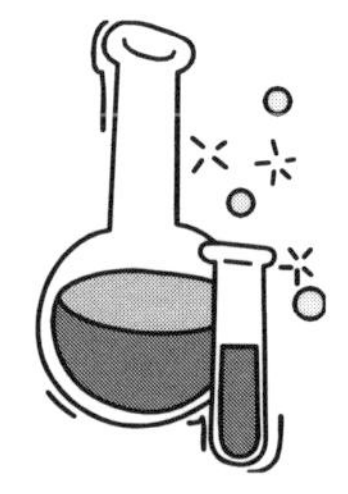

Safety Precautions: None

Procedure:
1. Blow up two balloons to their maximum inflation point and then tie them off. Try to inflate the balloons so they are the same size.
2. Determine their circumference by wrapping a piece of string around each balloon and then measuring the piece of string.
3. Place one balloon in the freezer overnight. Leave the other balloon outside of the freezer, to serve as a control.
4. The next day, remove the balloon from the freezer and immediately determine its circumference. Compare its circumference to the balloon that was not put into the freezer.

Explanation: The balloon that was placed in the freezer overnight should be considerably smaller than the balloon that was not. Charles's Law states that as the temperature of a gas decreases, its volume decreases. As the temperature falls, the molecules of the gas slow down. Therefore, the collisions of the molecules against the inside walls of the balloon are less frequent, and the collisions that do occur are much less forceful. This causes the balloon to decrease in volume

This effect is very noticeable with a helium-filled balloon that is taken outside on a cold winter day. The volume can decrease so much that the balloon will no longer remain buoyant.

Gases, Liquids, and Solids – Experiment # 13:
THE EXPANDING MARSHMALLOW

Objective: To cause a marshmallow to expand by placing it in a syringe.

Materials:

- Syringe with cap
- Marshmallow

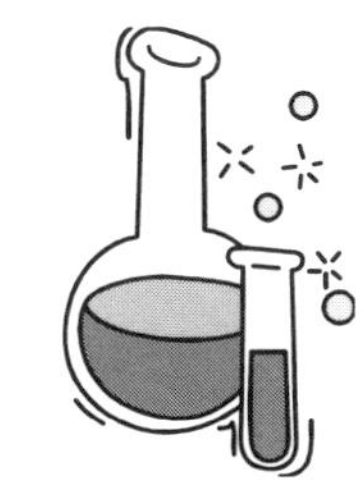

Safety Precautions: None

Procedure:

1. Remove the plunger and cap from the syringe.
2. Place a marshmallow in the syringe. (The large diameter syringes work the best, because you can place a full-sized marshmallow inside. If you have a smaller syringe, mini-marshmallows will generally fit inside.)
3. Insert the plunger, and press down until the plunger is just above the marshmallow.
4. Replace the cap. Lift the plunger up as far as you can. The marshmallow will noticeably expand.
5. Remove the cap. As air rushes back into the syringe, the marshmallow shrinks noticeably.

Explanation: As the plunger is lifted, a region of reduced pressure is created in the syringe above the marshmallow. The air within the marshmallow is then able to escape, since the air pressure opposing it is less. This provides a good demonstration of Boyle's Law, which states that as the pressure around a gas decreases, its volume increases. When air is introduced back into the syringe, the marshmallow shrinks. As the pressure on a gas increases, its volume decreases.

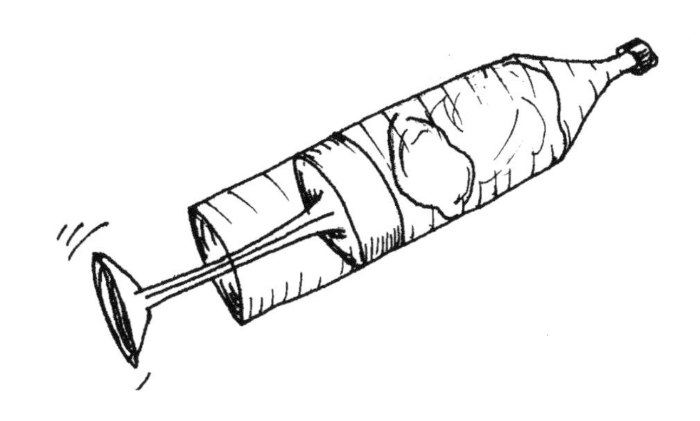

Gases, Liquids, and Solids – Experiment # 14:

MELTING A PENNY

Objective: To demonstrate the difference in melting points between two different metals.

Materials:

- Propane torch
- Matches
- Two pennies: one before 1982 and one after 1982
- Tongs

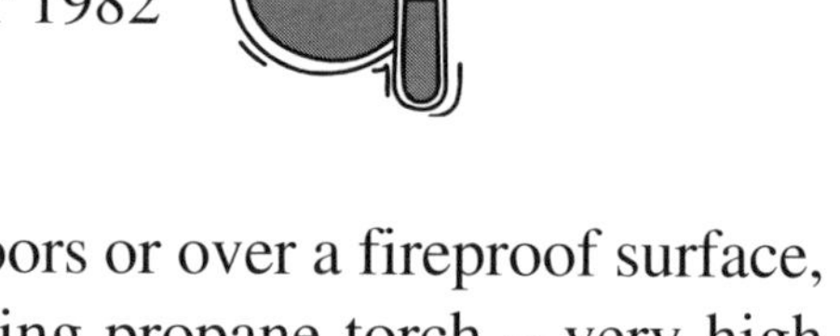

Safety Precautions: Do this experiment outdoors or over a fireproof surface, such as concrete. Exercise extreme care when using propane torch – very high temperatures are reached. Wear safety goggles.

Procedure:

1. Hold the pre-1982 penny with tongs and heat with the propane torch for several minutes.
2. Douse with water, then observe.
3. Repeat with the post-1982 penny. Observe. Make sure this is done outdoors or over a fireproof surface.

Explanation: The pre-1982 penny will still be essentially intact. But the newer penny will have melted into an indistinguishable glob of silver-colored metal. Before 1982, pennies were composed primarily of copper. The melting point of copper is approximately 1083°C, too high to be melted by the flame of the propane torch.

There is actually more than a penny's worth of copper in older pennies. Since many people were hoarding these pennies, after 1982 the U.S. mint began making pennies primarily from zinc, with just a thin copper coating. It costs much less to make a penny from zinc than from copper.

The melting point of zinc is approximately 419°C, and is easily melted by the propane torch. In the year 1982, you can find pennies that are made from either copper or zinc. The mass of newer pennies is also less that of older pennies, since zinc is less dense than copper.

Gases, Liquids, and Solids – Experiment # 15:
MELTING A MILK JUG

Objective: To discover why certain types of plastics become clear upon heating.

Materials:

- Candle
- Matches
- Scissors
- Plastic milk jug
- Tongs

Safety Precautions: Wear safety goggles. Do outside or in a well-ventilated area. Fumes given off may be toxic. Exercise caution around open flames. As the plastic melts, it may drip. Have a pan of water nearby to place the hot plastic into when finished.

Procedure:

1. Using scissors, cut out a section of milk jug approximately 5 x 5 cm. The plastic should be translucent.
2. Using the pair of tongs, hold over the candle flame until the plastic begins to melt. Once melting begins, the plastic will become transparent.
3. As soon as the plastic begins to drip, extinguish in a pan of water.

Explanation: As the plastic is heated, the molecules move faster and thus spread farther apart from one other. Since the molecules are farther apart, light is able to pass through, making the plastic transparent.

To visualize this, hold the fingers of your hand over a flashlight, making sure your fingers are together. Imagine that your fingers are molecules that prevent most of the light from passing through. As you spread your fingers out, then more light can pass through. This is what happens as the molecules of the plastic are heated and begin to spread out. It is only as the molecules move apart that the plastic can make the transition from being translucent to transparent.

Gases, Liquids, and Solids – Experiment # 16:

THE JOULE-THOMPSON EFFECT

Objective: To demonstrate the Joule-Thomson Effect using a balloon.

Materials:
- Balloon

Safety Precautions: None

Procedure:
1. Inflate a balloon with your lungs. Quickly feel the outside surface of the balloon and note its temperature. It should feel warm.
2. Allow the balloon to deflate. Quickly feel the outside of the balloon. It should feel cool.

Explanation: The Joule-Thomson effect states that as a gas is released through a small opening, its temperature drops. That is because it requires energy to force the gas through a small opening, and this removal of energy results in a decrease in temperature. The opposite is also true – as a gas is compressed, its temperature increases. This is because we are adding energy when we force air into the balloon. When a balloon is inflated energy is added and the temperature rises. When the air is let out, energy is being taken away, and the temperature drops. The same thing can be observed when a bicycle or car tire is inflated. Upon inflation, energy is added and the tire will feel warm. If deflated, the air will feel cool.

Gases, Liquids, and Solids – Experiment # 17:
EXAMINING CONDENSATION

Objective: To discover why condensation forms on objects that experience a rapid temperature decrease.

Materials:
- Glass drinking glass
- Freezer

Safety Precautions: None

Procedure:
1. Place a drinking glass into the freezer for about an hour.
2. Remove and observe.

Explanation: Condensation will immediately form on the glass when it is removed from the freezer. As the cold glass is brought into a warmer room, the glass immediately cools the air that is directly around it. This causes the water vapor that is in the air to condense into droplets of liquid water, which is what we see on the glass. This happens because the molecules that comprise warm air are moving faster than the molecules of cold air, therefore warm air can "hold" more water molecules than cold air. Water molecules in warm air are colliding so rapidly that they are less likely to coalesce into water droplets. As the temperature of the air (and thus the speed of the water molecules in the air) decreases, the collisions between water molecules become less violent. As a result, the water vapor molecules are more likely to combine and form tiny visible water droplets, which we observe as condensation, or liquid water. This process illustrates why dew forms at night. Since it gets cooler at night, some of the water vapor from the daytime warmer air will condense as dew.

Gases, Liquids, and Solids – Experiment # 18:
HEATING WATER IN A BALLOON

Objective: To demonstrate that water will keep a balloon cool enough to prevent a flame from popping it.

Materials:

- Several balloons
- Candle
- Matches

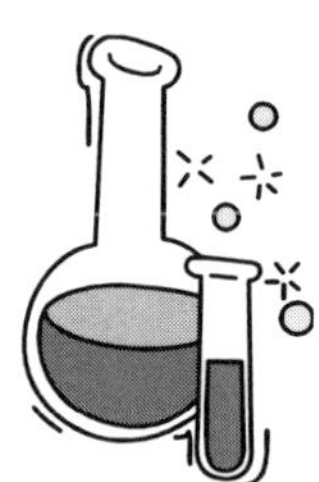

Safety Precautions: Exercise caution when using matches.

Procedure:

1. Blow up a balloon and tie it off.
2. Hold the balloon over a candle flame and note how long it takes to pop.
3. Now fill another balloon with water and tie it off.
4. Hold the water-filled balloon over the candle flame and observe. The balloon should not break.

Explanation: With the first balloon, the flame simply burns the rubber, causing the balloon to rupture. With the second balloon, the water conducts the heat away from the balloon, thus keeping the surface of the balloon cool and preventing it from popping. Water is a much better conductor of heat than is rubber. Rubber is an excellent insulator, which means it transfers energy very poorly. But if the rubber is in contact with the water, then the water is able to conduct the heat from the candle away from the rubber, preventing the balloon from popping.

Water has a high specific heat, which enables it to absorb a relatively large amount of energy before its temperature rises. The specific heat of water is 4.18 J/g°C. This means that if 1 gram of water absorbs 4.18 Joules (1 calorie) of energy, its temperature will rise 1°C. Since the specific heat of water is much larger than that of rubber, water can absorb more energy before its temperature rises.

Related Experiment: The same effect can be accomplished by filling a paper cup with water and heating it to boiling! The water conducts heat away from the cup, preventing it from burning.

Gases, Liquids, and Solids – Experiment # 19:
BURNING A DOLLAR BILL

Objective: To engulf a dollar bill in flames without harming it.

Materials:

- Dollar bill
- 70% isopropyl alcohol
- Cup
- Matches
- Tongs

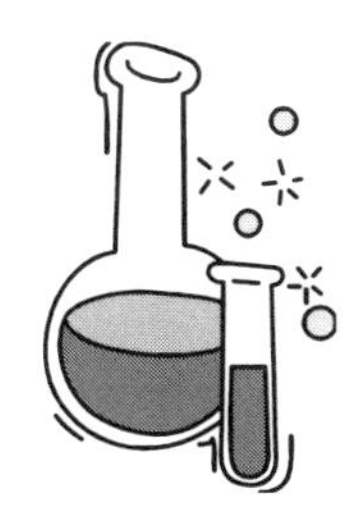

Safety Precautions: Wear safety goggles. Alcohol is extremely flammable. Keep away from flames. Never pour alcohol on an open flame. Always exercise caution with matches. Isopropyl alcohol can be fatal if swallowed. Do not inhale vapors.

Procedure:

1. Pour 30 mL of water in a cup.
2. To the water, add 60 mL of rubbing alcohol.
3. Thoroughly immerse the dollar bill in the alcohol-water mixture.
4. Hold the dollar bill with tongs and light with a match.
5. The dollar bill will burst into flames, but after the flames go out, the dollar bill will not be burned.

Explanation: If the dollar bill was placed in pure alcohol and then lit, the bill would be consumed. Yet when placed in a mixture of alcohol and water and then lit, the bill is untouched. As the alcohol burns, the water evaporates. Evaporation absorbs energy from its surroundings. As a result, the dollar bill never becomes hot enough to burn. It is the cooling effect of evaporation that prevents combustion of the dollar bill. This is why we perspire when overheated – the evaporation of water on our skin keeps us cool.

Gases, Liquids, and Solids – Experiment # 20:

REMOVING THE WATER FROM A CRYSTAL

Objective: To remove the water from a crystal by heating.

Materials:

- Microwave oven
- Copper(II) sulfate crystals (available in hardware stores as a root killer)
- Paper towels
- Microwaveable plate
- Sensitive balance

Safety Precautions: Copper(II) sulfate crystals are highly toxic if ingested – exercise caution. Use microwave only under adult supervision.

Procedure:

1. Using a sensitive balance, record the mass of some copper(II) sulfate crystals. About 5-10 grams should suffice.
2. Place these crystals on some paper towels on a microwaveable plate. Note the deep blue color of the crystals.
3. Place in the microwave oven. Turn the microwave on high and heat for 5-10 minutes.
4. When the crystals have turned white, remove.
5. Record the mass of the crystals now.

Explanation: The copper(II) sulfate crystals are an example of a hydrate – a substance that has molecules of water adhering to each of its own formula units. (A formula unit is the most basic unit of an ionic compound. Since the formula for copper(II) sulfate is $CuSO_4$, a single particle of $CuSO_4$ comprises one formula unit.) The copper(II) sulfate is an example of a pentahydrate. For every formula unit of copper sulfate, there are five molecules of water attached to it. The complete formula for copper sulfate is $CuSO_4{\cdot}5H_2O$.

Heating the crystals causes the water molecules to gain more energy and leave the crystal, being released as water vapor. When the water has left the crystals, they become white in color. This is known as their anhydrous form, which is without water. If the crystals are allowed to remain exposed to the air for several days, they will regain their blue color, since they will attract water from the atmosphere. The mass of the crystals after heating will be noticeably less. If all the water is driven out, it will decrease the mass by 36%.

If water is added to the white anhydrous crystals, they will instantly regain their deep blue color. They will also become quite hot, as the hydration of these crystals are highly exothermic.

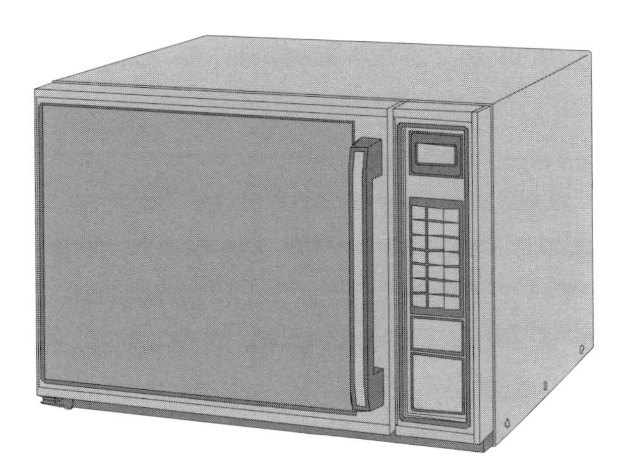

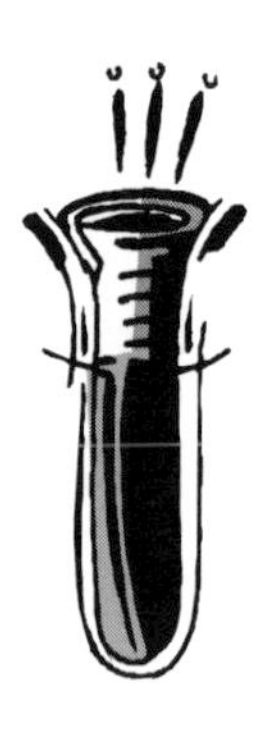

CHAPTER 8
CHEMICAL REACTIONS

Chemical reactions represent the heart and soul of chemistry. The bonding together of two or more atoms to produce a compound with completely different properties than either of the constituent elements is one of the marvels of chemistry.

Consider common table salt, or sodium chloride. It is composed of sodium and chlorine. Sodium is a silvery metal that bursts into flames when added to water. Chlorine is a poisonous yellow-green gas that can be deadly if inhaled. Together, they form a white compound that we probably add to our food every day.

Water is composed of two gases – hydrogen and oxygen. Hydrogen is an explosive gas. Combustion could not occur without oxygen. Yet together, they combine to form water, which is used to put out fires.

The incredible world of chemical reactions awaits you on the following pages . . .

Chemical Reactions – Experiment # 1:

CREATING RUST

Objective: To create rust instantly by the combination of several substances.

Materials:

- Vinegar (dilute acetic acid – $HC_2H_3O_2$)
- Household bleach (5% solution of sodium hypochlorite – NaOCl)
- Steel wool
- Plastic cup

Safety Precautions: This experiment releases chlorine gas fumes – it must be done outdoors or in a well-ventilated area. Bleach is very caustic – exercise caution. Do not inhale vapors or get it on your skin. Be careful not to spill any on your clothes. Wear safety goggles.

Procedure:

1. Add 30 mL of vinegar to the cup.
2. Add an equal amount of bleach to the same cup.
3. Add a small piece of steel wool the size of a large marble.
4. Observe for about 30 minutes. You will notice a rust-colored substance forming on the steel wool.

Explanation: Rust is the common name given to compounds produced by the oxidation of iron. They usually have a reddish-brown appearance. Oxidation occurs when a substance loses electrons, which occurs in this experiment. When an acid is added to bleach, chlorine gas (Cl_2) is released. The Cl_2 then reacts with the iron in the steel wool to produce iron(III) chloride, which is a reddish-brown color. Since vinegar is a weak acid, only a small amount of chlorine gas is released. The balanced chemical equation for the production of chlorine gas is as follows:

$$\underset{\text{(vinegar)}}{4HC_2H_3O_{2(aq)}} + \underset{\text{(bleach)}}{5NaOCl_{(aq)}} \Rightarrow \underset{\text{(chlorine)}}{2Cl_{2(g)}} + \underset{\text{(sodium chlorate)}}{NaClO_{3(aq)}} + \underset{\text{(sodium acetate)}}{4NaC_2H_3O_{2(aq)}} + 2H_2O$$

The chlorine gas produced then reacts with the iron in the steel wool to form iron(III) chloride. The balanced chemical reaction is as follows:

$$2Fe_{(s)} + 3Cl_{2(g)} \Rightarrow 2FeCl_{3(s)}$$

In the above equation, iron is oxidized, since it loses electrons. Its charge is increased from 0 in elemental iron to 3+ in iron(III) chloride. The Cl_2 is reduced in the above reaction. Its charge is reduced from 0 to 1-. Since the iron(III) chloride is produced as a result of the oxidation of iron, it is classified as rust.

The combination of the above three substances is an excellent example of a chemical change or a chemical reaction – which is characterized by the creation of a new substance.

Chemical Reactions – Experiment # 2:
TO RUST OR NOT?

Objective: To determine which substances rust and which do not.

Materials:

- Iron nails
- Galvanized (zinc-plated) nails
- Penny
- Aluminum foil
- Other types of metals
- Paper towels
- Magnet

Safety Precautions: None

Procedure:

1. Place the metals on a paper towel, making sure that none are touching.
2. Soak the towel thoroughly with water, and place another soaked towel on top of it.
3. Allow to remain undisturbed for several days, keeping the towels moist.
4. At the end of this time, carefully observe each. You should notice rust only on the metals composed of iron.
5. To verify whether or not a compound is composed of iron, test with a magnet. Iron is one of the few metals that are attracted to a magnet.

Explanation: Rust is a general term covering several compounds of iron, often some form of iron oxide. Therefore, the only substances capable of rusting are those composed of iron. Other metals may undergo other types of corrosion, but it cannot be considered rust unless it involves iron. Oxygen and water are also necessary for rusting to occur. The balanced chemical equation for the formation of rust is as follows:

$$4Fe_{(s)} + 3O_{2(g)} + 3H_2O_{(l)} \Rightarrow 2Fe_2O_3 \cdot 3H_2O_{(s)}$$

Rust is commonly written as just Fe_2O_3, with the water omitted. But from the equation above, it can be seen that rust is actually an example of a hydrate. A hydrate has water molecules bound to the ionic solid in a fixed ratio.

Chemical Reactions – Experiment # 3:

WHAT DETERMINES THE RATE OF RUSTING?

Objective: To determine the effect of salt on the rusting of iron.

Materials:

- Iron nails
- Paper towels
- Salt

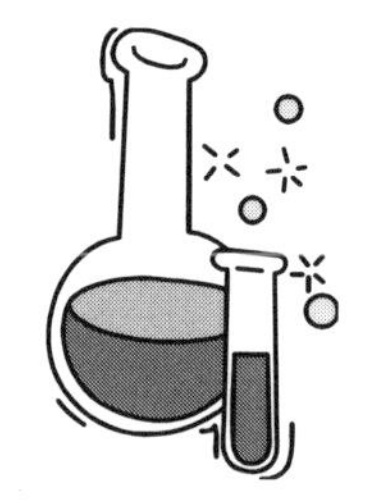

Safety Precautions: None

Procedure:

1. Place several iron nails on a wet paper towel, and cover with another wet paper towel.
2. On a separate paper towel repeat the above, except sprinkle salt liberally on the nails and paper towels.
3. Allow both to remain undisturbed for several days. At the end of this period, observe and compare the amount of rusting for each.
4. Repeat the above experiment with different substances, to determine their effect on rusting.

Explanation: The nails with salt added should have experienced a much greater degree of rust than those with no salt. Salt acts as a catalyst, which is a substance that speeds up the rate of a reaction without actually participating in the reaction.

Cars in the northern regions of the United States rust at a much greater rate than those in the southern parts, due to the large amount of salt used on the roads to melt the ice in the wintertime. The salt greatly speeds up the rate of rusting on the iron bodies of automobiles.

Chemical Reactions – Experiment # 4:

HOW TO REMOVE RUST

Objective: To determine which substances are most effective at removing rust.

Materials:

- Vinegar
- Lemon juice
- Coca-Cola
- Ammonia
- Plastic cups
- Rusty nails or bolts
- Plastic wrap
- Rubber bands
- pH paper (available from a pet shop or pool supply store)

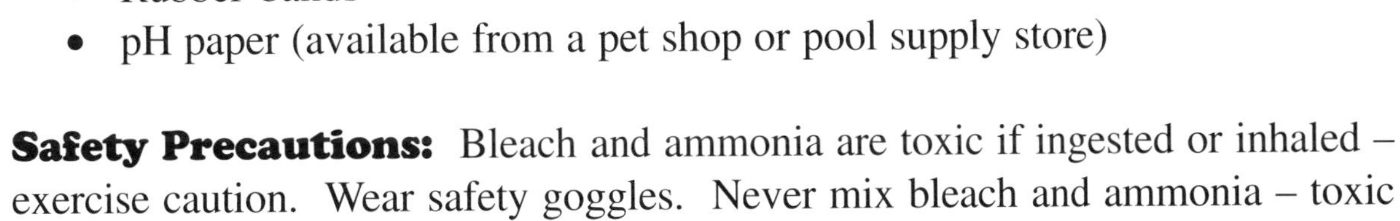

Safety Precautions: Bleach and ammonia are toxic if ingested or inhaled – exercise caution. Wear safety goggles. Never mix bleach and ammonia – toxic fumes are released.

Procedure:

1. Pour a quarter-cup of each of the above liquids.
2. Drop a rusty nail in each. Cover each with plastic wrap, and secure with a rubber band. Allow to remain undisturbed for one week.
3. Remove each nail and observe for signs of rust.
4. Test the pH of each substance with pH paper, to determine if this has an effect on its rust-removing abilities.

Explanation: You should notice that the acidic substances – Coca-Cola, lemon juice, and vinegar – were the most effective at removing rust. This can be verified by testing with pH paper, if available. Acidic substances have a pH less than 7 at 25°C. Most other substances either have no effect or may have even hastened the rusting process. Acids remove rust by chemically reacting with it.

Chemical Reactions – Experiment # 5:

TURNING PENNIES GREEN

Objective: To observe the formation of a new compound on the surface of a penny due to its reaction with vinegar.

Materials:

- Several shiny pennies
- Paper towels
- Vinegar
- Salt

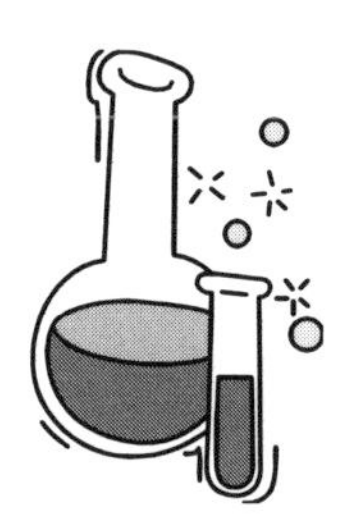

Safety Precautions: None

Procedure:

1. Soak a paper towel in vinegar.
2. Place several pennies on one half of the towel, and then fold the other half over.
3. Repeat the above steps on a separate vinegar-soaked paper towel, sprinkling salt on each penny.
4. Allow both to remain undisturbed overnight. The next day, you should observe a greenish coating on each penny.

Explanation: The vinegar, or acetic acid, reacts with the copper of the pennies to form a new compound known as copper(I) acetate. This substance is responsible for the green color on the pennies. The salt acts as a catalyst, causing more copper(I) acetate to form on each penny. A catalyst speeds up the rate of a reaction. The balanced chemical equation is as follows:

$$\underset{\text{copper}}{2Cu_{(s)}} + \underset{\text{vinegar}}{2HC_2H_3O_{2(aq)}} \Rightarrow \underset{\text{copper(I) acetate}}{2CuC_2H_3O_{2(s)}} + \underset{\text{hydrogen}}{H_{2(g)}}$$

This is an oxidation-reduction, or redox reaction. A redox reaction involves the transfer of electrons. In this reaction, the copper is oxidized, since its charge increases from a 0 charge in its pure form to a 1+ charge in copper(I) acetate. The hydrogen is reduced, with its charge decreasing from 1+ to 0. Since the copper and hydrogen exchange places, this experiment is also a good example of a single displacement reaction.

Chemical Reactions – Experiment # 6:
EXCHANGING COPPER FOR IRON

Objective: To create a copper coating on an iron nail.

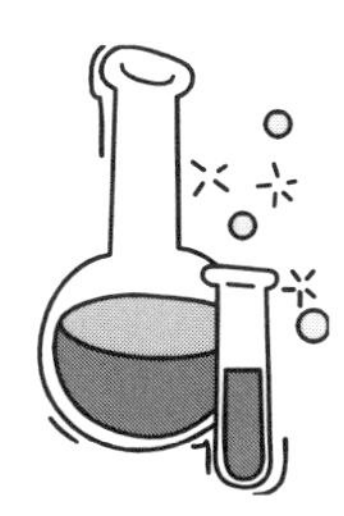

Materials:

- Copper(II) sulfate (available in drug stores and hardware stores – often sold as a root killer)
- Non-galvanized iron nail
- Plastic cup

Safety Precautions: Copper(II) sulfate is extremely toxic if ingested. Wash hands after using.

Procedure:

1. Make a solution of copper(II) sulfate by mixing one teaspoon of granular copper(II) sulfate with about a half-cup of water. Stir thoroughly until dissolved.
2. Drop an iron nail into the cup of copper(II) sulfate solution, making sure the nail is completely submerged.
3. After 45 minutes, remove the nail and observe. It will be covered with a thick coating of copper, which immediately darkens upon contact with the air.

Explanation: This is an example of a single displacement reaction – where one element is exchanged with another. In this reaction, the iron in the nail trades places with the copper in the copper(II) sulfate, because iron is a much more active metal than copper. In other words, iron loses electrons much more readily than does copper. Therefore the copper is deposited on the nail and the liquid solution becomes iron(II) sulfate. The balanced chemical equation is as follows:

$$\underset{\text{copper(II) sulfate}}{CuSO_{4(aq)}} + Fe_{(s)} \Rightarrow \underset{\text{iron(II) sulfate}}{FeSO_{4(aq)}} + Cu_{(s)}$$

Upon exposure to the air, the copper reacts with oxygen to form copper(II) oxide, which is black in color. The reaction is as follows:

$$2Cu_{(s)} + O_{2(g)} \Rightarrow 2CuO_{(s)}$$

Chemical Reactions – Experiment # 7:
TURNING AMMONIA BLUE

Objective: To form a new substance by reacting copper with ammonia.

Materials:

- Penny
- Ammonia
- Plastic cup
- Plastic wrap
- Rubber band

Safety Precautions: Ammonia is toxic if inhaled or ingested. Exercise caution. Wear safety goggles.

Procedure:

1. Place a penny in a small plastic cup.
2. Add enough ammonia to completely cover the penny. Cover with plastic wrap and secure with a rubber band.
3. Observe daily for a one week period.

Explanation: The ammonia and copper react to form a new substance known as the tetraamine copper(II) ion. This is an example of a complex ion. A complex ion is a charged substance composed of a metal ion surrounded by a ligand. A ligand is a substance that covalently bonds to the metal ion due to having an available pair of free electrons. Complex ions are often characterized by their intense colors. In this case, a deep blue color is formed. The formula for this substance is $[Cu(NH_3)_4]_4^{2+}$. The balanced chemical equation is as follows:

$$4Cu_{(s)} + 16NH_{3(aq)} \Rightarrow [Cu(NH_3)_4]_4^{2+}$$

Chemical Reactions – Experiment # 8:

THE DISAPPEARING ALUMINUM FOIL

Objective: To make aluminum foil "disappear" by reacting with hydrochloric acid.

Materials:

- Muriatic acid (available in hardware stores)
- Small glass jar
- Aluminum foil

Safety Precautions: Muriatic acid contains hydrochloric acid, which is extremely corrosive to the eyes and skin. Always wear safety goggles when dealing with acids. Do this experiment outdoors or in a well-ventilated area, since very irritating fumes are produced. Do not replace the lid on the container after adding the foil, since the gas given off may build up enough pressure to cause the container to burst if sealed.

Procedure:

1. Pour about 30 mL of muriatic acid in a glass jar.
2. Place several small pieces of aluminum foil into the jar. Observe.

Explanation: The aluminum reacts with the hydrochloric acid to yield aluminum chloride and hydrogen gas. This is why the aluminum is disappearing – it is reacting with the acid to form a brand new substance. The new substance that forms is an aqueous solution of aluminum chloride, which appears invisible since it is dissolved in water. The stronger the concentration of acid, the quicker the aluminum foil will react. The balanced chemical equation is as follows:

$$2\,Al_{(s)} + 6HCl_{(aq)} \Rightarrow 2\,AlCl_{3(aq)} + 3H_{2(g)}$$

Chemical Reactions – Experiment # 9:

EXTRACTING COPPER FROM COPPER SULFATE

Objective: To demonstrate that aluminum foil can be used to remove the copper from copper sulfate.

Materials:

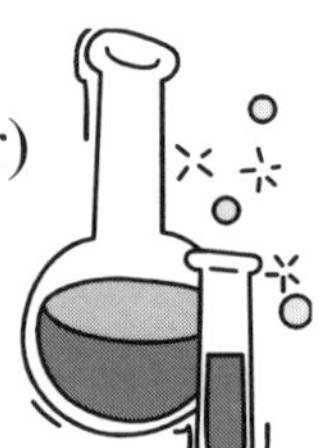

- Copper(II) sulfate (available in hardware stores as a root killer)
- Plastic cup
- Aluminum foil
- Table salt (sodium chloride)

Safety Precautions: Copper(II) sulfate is toxic if ingested. The copper that is extracted during this experiment should not be touched – it is very toxic if ingested. Place in a sealed jar when finished and dispose of at a hazardous waste facility. Wash hands thoroughly after this experiment.

Procedure:

1. Make a solution of copper(II) sulfate by adding a teaspoon of copper(II) sulfate to about a half-cup of water. Mix thoroughly until dissolved.
2. Cut out a strip of aluminum foil approximately four inches by four inches and roll loosely.
3. Place the foil in the cup of copper(II) sulfate solution and observe for several minutes.
4. Now add several teaspoons of table salt and stir. You will immediately notice a chemical reaction occurring.

Explanation: Several different chemical reactions occur in this experiment. The reaction most easily observed is a single displacement reaction in which the aluminum and the copper trade places. Pure copper is thus produced, and the liquid formed is aluminum sulfate. The balanced chemical equation for this reaction is as follows:

$$\underset{\text{copper(II) sulfate}}{3CuSO_{4(aq)}} + 2Al_{(s)} \Rightarrow \underset{\text{aluminum sulfate}}{Al_2(SO_4)_{3(aq)}} + 3Cu_{(s)}$$

However, this reaction will not occur until sodium chloride is added. Since aluminum is readily oxidized in air, the aluminum is coated with a layer of aluminum oxide (Al_2O_3), which prevents the copper ions from reacting with it. The chloride ion from the salt will react with this aluminum oxide coating, exposing the pure aluminum underneath and allowing the copper to react with it.

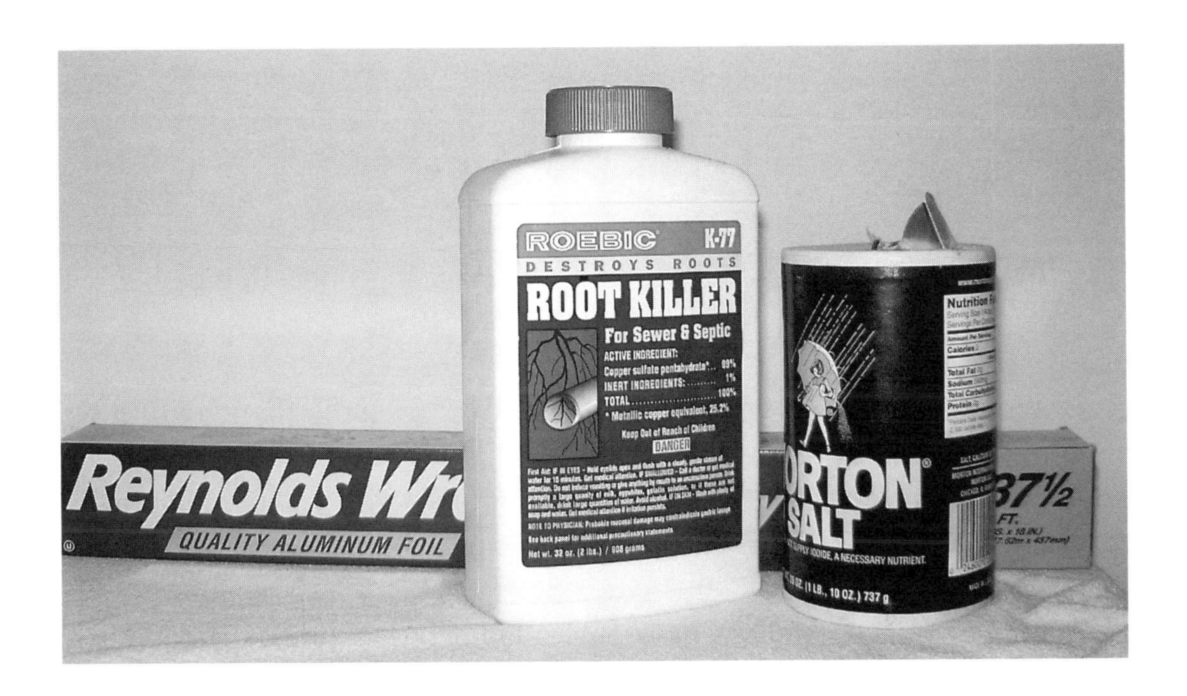

Chemical Reactions – Experiment # 10:

BURNING A SUGAR CUBE

Objective: To demonstrate that a catalyst must be used in order to burn a sugar cube.

Materials:

- Sugar cube
- Tongs
- Candle
- Cigarette ashes

Safety Precautions: Exercise caution when using open flames.

Procedure:

1. Holding a sugar cube with tongs, hold it over a candle flame until it begins to melt. It will not burn.
2. Now take another sugar cube and thoroughly cover it with cigarette ashes. This cube will now burn, and should continue to burn for a couple of minutes.

Explanation: The cigarette ashes contain lanthanide oxides. The lanthanides are known as the rare earth metals, and comprise elements 58-71 on the periodic table of the elements. In this experiment the lanthanide oxides within cigarette ashes act as a catalyst, allowing the sugar cube to combust instead of melt. A catalyst is a substance that speeds up the rate of a chemical reaction, but is not consumed or changed during the reaction. As sugar combusts, the carbon is released, which will be black. The hydrogen and oxygen will be released as water vapor. Note that the ratio of hydrogen and oxygen in sugar is the same as that in water – a 2:1 ratio. The balanced chemical equation for the combustion of sugar is as follows:

$$C_{12}H_{22}O_{11(s)} \Rightarrow 12C_{(s)} + 11H_2O_{(g)}$$

Chemical Reactions – Experiment # 11:

MAKING YOUR PENNIES CLEAN

Objective: To discover which substances are the most effective at cleaning pennies.

Materials:

- Several discolored pennies
- Plastic cups
- Ketchup
- Vinegar
- Lemon juice
- Coca-Cola
- Plastic wrap
- Rubber bands

Safety Precautions: Always read the label before using any household substance, and exercise proper precautions accordingly.

Procedure:

1. Place a penny in each of several plastic cups.
2. Add enough of each of the above substances to completely cover each penny. Cover each with plastic wrap and secure with a rubber band.
3. Label each accordingly.
4. Make daily observations for one week.

Explanation: The most effective copper cleaners will be those that are acidic – ketchup, carbonated beverages, vinegar, and lemon juice. Numerous types of corrosion appear on copper, due to reactions with compounds in the air or other substances the copper may come into contact with. The dark coating on the surface of pennies is often composed of copper(II) oxide or cupric oxide, formed from the oxidation of the copper with oxygen in the air. Acids react with the copper(II) oxide, as well as other types of corrosion, and remove it, thus forming clean pennies.

Chemical Reactions – Experiment # 12:
CREATING COPPER OXIDE

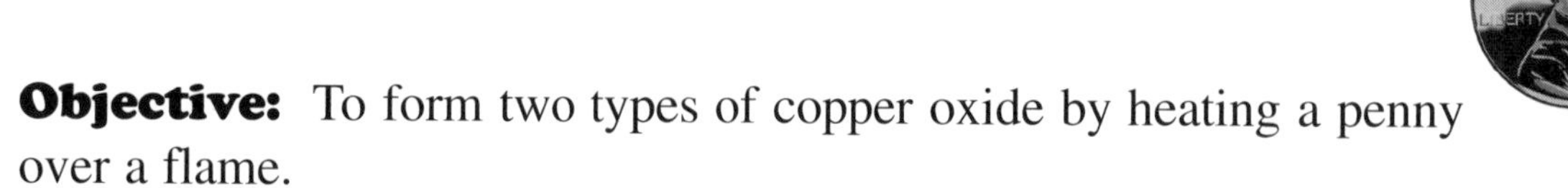

Objective: To form two types of copper oxide by heating a penny over a flame.

Materials:

- Propane torch or gas stove
- Pre-1982 penny
- Tongs or tweezers

Safety Precautions: Always exercise caution around open flames. Use an older penny since newer pennies will melt. Wear safety goggles.

Procedure:

1. Using tongs, hold a penny in a flame for 30 seconds to one minute. You should observe a distinct reddish color.
2. Continue to heat for five minutes. The penny will turn completely gray.

Explanation: The heating of the penny results in the combining of oxygen with the copper. This is a combustion reaction. Initial heating will yield a reddish color, which is copper(I) oxide, or cuprous oxide.

Additional heating causes further oxidation of the copper to copper(II), resulting in copper(II) oxide, or cupric oxide. We have thus formed two compounds of copper by the combining of copper with oxygen. The copper(I) is a reddish color, and the copper(II) is a gray-black color. The balanced chemical equations are as follows:

$$4Cu_{(s)} + O_{2(g)} \Rightarrow 2Cu_2O_{(s)}$$

$$2Cu_2O_{(s)} + O_{2(g)} \Rightarrow 4CuO_{(s)}$$

Chemical Reactions – Experiment # 13:

GENERATION OF OXYGEN

Objective: To produce oxygen gas using yeast.

Materials:

- 3% hydrogen peroxide
- Active, dry yeast
- Glass jar
- Match
- Wooden coffee stirrer

Safety Precautions: Exercise caution when using open flames. Hydrogen peroxide is toxic if ingested.

Procedure:

1. Add about 30 mL of hydrogen peroxide to a glass jar.
2. Add about ½ teaspoon of yeast, and gently swirl. After a few minutes, bubbles will form.
3. After bubbles have formed, light a wooden coffee stirrer.
4. Blow out the flame, so only glowing embers remain.
5. Lower the glowing coffee stirrer down into the jar. It will burst into flames. Blow out the flames and again insert into the jar.

Explanation: Yeast are one-celled fungi that secrete enzymes which act as catalysts. These catalysts break down hydrogen peroxide into water and oxygen. The balanced chemical equation is as follows:

$$2H_2O_{2(l)} \Rightarrow 2H_2O_{(l)} + O_{2(g)}$$

Oxygen supports the combustion of other materials, but does not itself burn. In the presence of oxygen the glowing embers therefore burst into flames. This explains why blowing on the embers of a campfire can cause them to ignite.

Chemical Reactions - Experiment # 14:

ANOTHER METHOD TO GENERATE OXYGEN

Objective: To generate oxygen using manganese dioxide as a catalyst.

Materials:

- 3% hydrogen peroxide
- Depleted alkaline flashlight battery
- Hacksaw
- Glass jar
- Matches
- Wooden coffee stirrer

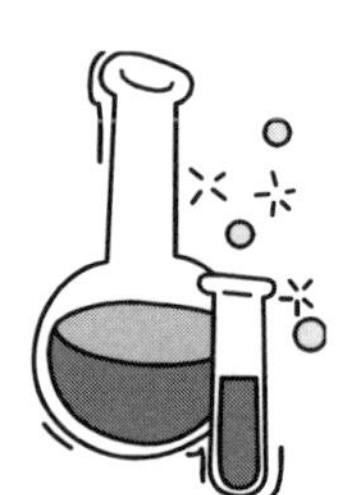

Safety Precautions: Exercise caution when sawing battery in half. Hydrogen peroxide is toxic if ingested. Wear safety goggles.

Procedure:

1. Add 30 mL of hydrogen peroxide to a glass jar.
2. Carefully saw a depleted flashlight battery in half using a hacksaw.
3. Extract the black powder from the battery. This black material will be manganese dioxide (MnO_2), which is part of the electrolyte within a battery.
4. Add a small amount of the MnO_2 to the hydrogen peroxide in the jar. It should fizz.
5. Light a wooden coffee stirrer, then blow it out so it is still glowing.
6. Insert the glowing coffee stirrer into the jar. The glowing embers should burst into flames. Blow out the flames and repeat.

Explanation: The manganese dioxide acts as a catalyst, splitting the hydrogen peroxide into water and oxygen. Oxygen supports the combustion of other substances, causing the glowing embers to burst into flames. The balanced chemical equation is as follows:

$$2H_2O_{2(l)} \Rightarrow 2H_2O_{(l)} + O_{2(g)}$$

Since the manganese dioxide acts as a catalyst and is not used up in the reaction, it is not included in the balanced chemical equation.

Chemical Reactions – Experiment # 15:

CREATING MAGNESIUM HYDROXIDE

Objective: To create magnesium hydroxide by means of a chemical reaction.

Materials:

- Three transparent plastic cups
- Epsom salts (magnesium sulfate)
- Clear household ammonia

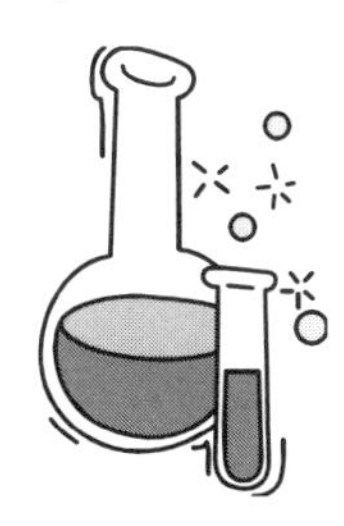

Safety Precautions: Ammonia is toxic if inhaled or ingested. Wear safety goggles.

Procedure:

1. Add three teaspoons of Epsom salts to a cup that is half-filled with water. Stir until the salt is completely dissolved. When completely dissolved, the solution will be clear, with no solids remaining on the bottom.
2. Add to the solution of Epsom salts a half-cup of clear household ammonia. Observe.
3. Allow to remain undisturbed for one week. Observe.

Explanation: The addition of ammonia to magnesium sulfate results in the formation of a precipitate, which causes the resulting solution to be cloudy. A precipitate is an insoluble solid that does not dissolve but rather precipitates, or settles, to the bottom of the container. After the end of one week, the precipitate should have mostly settled on the bottom. The precipitate that forms in this reaction is magnesium hydroxide, with a formula of $Mg(OH)_2$. The ammonia first of all reacts in water to form ammonium ions (NH_4^+) and hydroxide ions (OH^-). The balanced chemical equation for this reaction is as follows:

$$NH_{3(aq)} + H_2O_{(l)} \Rightarrow NH_4^+{}_{(aq)} + OH^-{}_{(aq)}$$

The hydroxide ions then react with the magnesium ions (Mg^{2+}) furnished by the Epsom salts to form the insoluble precipitate magnesium hydroxide, with a formula of $Mg(OH)_2$. The balanced chemical equation for this reaction is as follows:

$$Mg^{2+}_{(aq)} + 2OH^{-}_{(aq)} \Rightarrow Mg(OH)_{2(s)}$$

Magnesium hydroxide is the active ingredient in Milk of Magnesia, and since it is a base will relieve acid indigestion by neutralizing excess stomach acid.

Chemical Reactions – Experiment # 16:
FORMING IRON(III) CHLORIDE

Objective: To form iron(III) or ferric chloride by combining iron and chlorine.

Materials:

- Iron nail
- Plastic cup
- Household bleach
- Plastic wrap
- Rubber band

Safety Precautions: Bleach is very toxic if ingested. Do not allow bleach to contact your skin. Be careful not to spill it on your clothes. Wear safety goggles.

Procedure:

1. Place an iron nail in a plastic cup.
2. Add enough bleach to cover the nail.
3. Cover with plastic wrap and secure with a rubber band.
4. Allow to remain undisturbed for one week. Observe.

Explanation: The active ingredient in bleach is sodium hypochlorite (NaOCl). Household bleach releases chlorine gas (Cl_2), which is why bleach has a strong odor. The chlorine in the bleach reacts with the iron in the nail to form a thick brown substance that resembles rust. This substance is known as ferric chloride, or iron(III) chloride. The balanced chemical equation for this reaction is as follows:

$$2Fe_{(s)} + 3Cl_{2(aq)} \Rightarrow 2FeCl_{3(s)}$$

Bleach is an excellent oxidizer, meaning it readily removes electrons from other substances. The iron is therefore oxidized from its neutral state to a 3+ charge, due to the loss of 3 electrons. The chlorine is reduced from a charge of 0 to a 1- charge in the chloride ion.

Chemical Reactions – Experiment # 17:

REACTION IN A BAGGIE

Objective: To observe the signs that a chemical reaction has occurred.

Materials:

- Citric acid ($H_3C_6H_5O_7$)
- Baking soda (sodium bicarbonate – $NaHCO_3$)
- Quart freezer bag
- Film canister
- Red cabbage juice (or universal indicator solution if available)

Safety Precautions: Wear safety goggles – the bag may pop open. Do not shake the bag, and make sure the opening is pointed away from your eyes.

Procedure:

1. Place a half-teaspoon each of citric acid and baking soda in the bag.
2. Fill a film canister with red cabbage solution (see Acids and Bases – Experiment # 1). If unavailable, water can be used.
3. Lay the bag on a flat surface and carefully place the film canister upright within the bag, without spilling any liquid. Seal the bag, making sure that the contents of the canister do not spill.
4. Gently mix the contents of the bag. Make sure your safety goggles are on. Make as many observations as you can.

Explanation: The bag expands due to the production of carbon dioxide gas that is formed as a result of a chemical reaction between the baking soda and vinegar. The balanced chemical equation is as follows:

$$H_3C_6H_5O_{7(aq)} + NaHCO_{3(aq)} \Rightarrow Na_3C_6H_5O_{7(aq)} + H_2O_{(l)} + CO_{2(g)}$$

citric acid	sodium bicarbonate	sodium citrate	water	carbon dioxide

This reaction does not occur until water is added. Adding water dissolves the citric acid and baking soda, which greatly increases the surface area of these substances. As a result there are many more collisions that occur between reactant molecules and ions, which causes the reaction to occur.

You should have noticed a decrease in temperature as a result of this reaction as you felt the products within the bag. This reaction is endothermic, meaning it absorbs energy from its surroundings. As a result, the surroundings get colder.

The red cabbage juice will initially turn red due to the presence of the citric acid. The red cabbage juice changes colors depending on the pH of the solution. As the baking soda gradually neutralizes the citric acid, it will return closer to its original color. The final color of the red cabbage juice will depend on how much citric acid and baking soda was originally used.

This experiment provides four signs that a chemical reaction has occurred: the formation of a gas, a temperature change, a color change, and a pH change. Any of these signs by themselves may not necessarily indicate that a chemical reaction has occurred, but the simultaneous presence of all four of these signs is strong evidence that a chemical reaction has likely occurred.

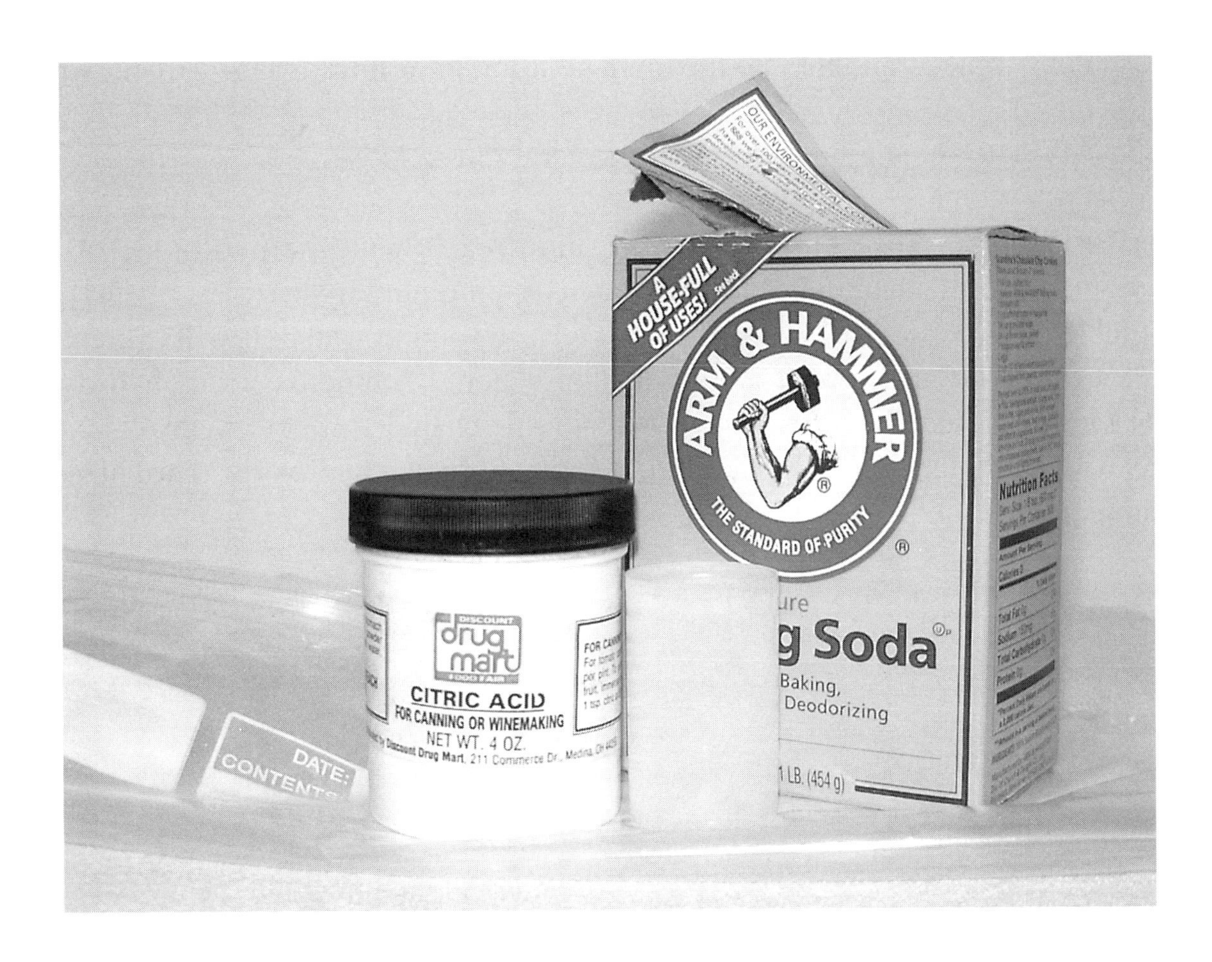

Chemical Reactions – Experiment # 18:

FORMING ALUMINUM HYDROXIDE

Objective: To form aluminum hydroxide, a solid, by the addition of two liquids.

Materials:

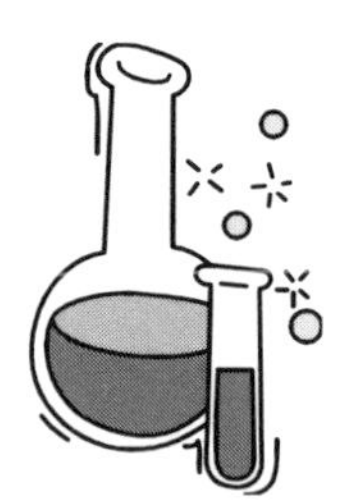

- Two transparent drinking glasses
- Alum (aluminum potassium sulfate)
- Household ammonia

Safety Precautions: Ammonia is toxic if inhaled or ingested. Exercise caution. Wear safety goggles.

Procedure:

1. Add two teaspoons of alum to a glass half-full of water. Stir thoroughly to dissolve.
2. Pour some ammonia into the alum solution until it turns into a thick, white mass.
3. Allow to remain undisturbed for several days. Observe.

Explanation: The aluminum of the alum reacts with the hydroxide of the ammonia to form a white precipitate known as aluminum hydroxide. $Al(OH)_3$ is insoluble in water, thus it is readily observed as a solid in water. It will eventually settle to the bottom, since it is denser than water. Ammonia forms hydroxide (OH^-) ions in water (see experiment # 15 in this chapter). These hydroxide ions react with aluminum ions from the alum to form aluminum hydroxide. The balanced chemical equation is as follows:

$$Al^{3+}_{(aq)} \quad + \quad 3OH^-_{(aq)} \quad \Rightarrow \quad Al(OH)_{3(s)}$$

Chemical Reactions – Experiment # 19:

LIGHT BLUE TO DARK BLUE

Objective: To change the color of copper(II) sulfate from light blue to dark blue.

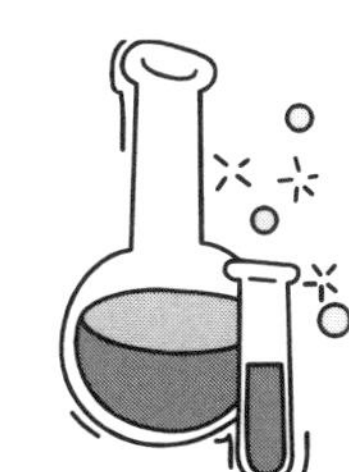

Materials:

- Copper(II) sulfate
- Transparent cup
- Household ammonia

Safety Precautions: Copper(II) sulfate and ammonia are both toxic if ingested. Do not inhale vapors of ammonia. Wear safety goggles.

Procedure:

1. Add just a few crystals of copper(II) sulfate to a cup of water. Stir until dissolved. The color should be a very faint light blue.
2. Pour a little ammonia into the solution. The color will change to deep blue.

Explanation: The ammonia reacts with the copper(II) sulfate to form what is known as a complex ion. Complex ions are often characterized by a very intense color. In this case, the complex ion $[Cu(NH3)4]_4^{2+}$ is produced, which is characterized by a deep blue color. This ion is known as the tetraamine copper(II) ion. (See experiment # 7 in this chapter for the balanced chemical equation.)

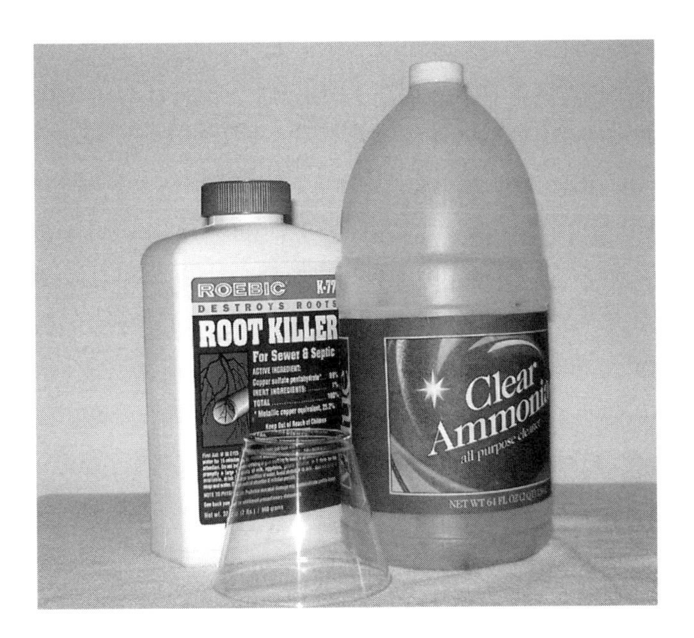

Chemical Reactions – Experiment # 20:

DOES TEMPERATURE AFFECT THE RATE OF A REACTION?

Objective: To determine whether or not temperature affects the rate of a chemical reaction.

Materials:

- Two light sticks (available from a toy or sporting goods store)
- Two microwaveable cups
- Microwave oven
- Ice cubes

Safety Precautions: Exercise caution with hot water.

Procedure:

1. Heat a cup of water to near boiling in the microwave oven.
2. Place ice cubes in another cup of water.
3. Activate both light sticks.
4. Place one light stick in the cold water and the other in the hot water. Turn off the lights and observe.

Explanation: The activation of the light stick is an example of chemiluminescence – a chemical reaction that produces light without heat. It is also known as cold light. This same type of reaction occurs in the firefly and some deep-sea creatures. Some types of fungi and bacteria also exhibit chemiluminescence. When the light stick is heated, its molecules move faster, increasing the rate at which they react. Therefore the light stick in the hot water should glow much more brightly. The cold water slows down the molecules, causing the reaction to proceed at a slower pace. This will cause the light stick to glow less brightly. If a glowing light stick is immersed in liquid nitrogen, which is very cold, it will cease glowing entirely.

Chemical Reactions – Experiment # 21:

THE DECOLORIZING EFFECTS OF BLEACH

Objective: To observe the effects of bleach on food coloring.

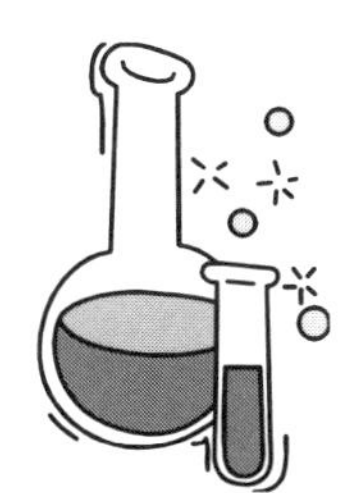

Materials:

- Clear plastic cup
- Household bleach
- Food coloring

Safety Precautions: Bleach is very toxic. Be careful not to spill bleach on your clothing. Wear safety goggles.

Procedure:

1. Fill the cup about halfway with bleach.
2. Add a few drops of food coloring. Stir to mix completely.
3. Observe for several minutes. The solution will become clear.

Explanation: Bleach chemically reacts with the food coloring to render it colorless. Household bleach is a solution of sodium hypochlorite, which is a good oxidizer. An oxidizer, or oxidizing agent, removes electrons from other substances. The bleach oxidizes the food coloring, changing it into a colorless substance. The decolorizing effects of bleach on fabrics and hair is due to a similar reaction.

Chemical Reactions – Experiment # 22:

WHY DOES HYDROGEN PEROXIDE FIZZ?

Objective: To determine why hydrogen peroxide fizzes when applied to a cut.

Materials:

- Raw potato
- Raw liver or ground beef
- 3% hydrogen peroxide
- Plastic cup

Safety Precautions: Hydrogen peroxide is poisonous. Exercise caution.

Procedure:

1. Slice a raw potato in half.
2. Pour some hydrogen peroxide on the potato. It should fizz.
3. Place a piece of raw liver or ground beef in a plastic cup. Pour some hydrogen peroxide over this. What happens?

Explanation: Potatoes contain an enzyme called catalase, which is responsible for breaking down hydrogen peroxide into water and oxygen. The release of oxygen bubbles produces the fizzing. Blood contains similar enzymes, which is why raw liver or ground beef, which both contain blood, also fizz when hydrogen peroxide is added. This explains why a cut will fizz when hydrogen peroxide is applied. Test other substances to see which fizzes when hydrogen peroxide is applied. The balanced chemical equation is as follows:

$$2H_2O_{2(l)} \Rightarrow 2H_2O_{(l)} + O_{2(g)}$$

Blood contains the catalase enzyme because hydrogen peroxide is naturally produced in our cells, and is extremely toxic. The catalase instantly breaks down this poison into water and oxygen.

Chemical Reactions – Experiment # 23:

HOW TO PREVENT AN APPLE FROM BROWNING

Objective: To discover a method to prevent an apple from browning after it has been exposed to the air.

Materials:

- Apple
- Vitamin C tablet
- Lemon juice
- Baking soda

Safety Precautions: None

Procedure:

1. Crush a vitamin C tablet into powder.
2. Slice an apple into quarters.
3. Leave the first piece untouched, to serve as a control.
4. Sprinkle some crushed vitamin C liberally on the second piece.
5. Pour some lemon juice on the third piece of the apple.
6. On the fourth piece, sprinkle some baking soda.
7. Observe every five minutes for one hour.

Explanation: The apple browns due to enzymes in the apple that react with oxygen in the air. The vitamin C acts as an inhibitor, reacting with these enzymes before they have a chance to react with oxygen in the air. Therefore the apple does not turn brown. The same result can be accomplished with lemon juice, which also contains vitamin C.

An inhibitor is the opposite of a catalyst. Catalysts speed up the rate of a chemical reaction, and an inhibitor slows down the rate of a reaction. The commercially available substance known as Fruit Fresh, which prevents cut fruit from browning, contains Vitamin C as its active ingredient.

The baking soda, which is a basic substance, speeds up the browning process. Generally, acidic substances will prevent fruit from browning, and basic substances will speed up the browning process. Vitamin C is ascorbic acid, and lemon juice contains citric acid. Experiment with other substances to determine their effect on the browning of fruit.

Chemical Reactions – Experiment # 24:
TESTING FOR STARCH

Objective: To determine which foods contain starch.

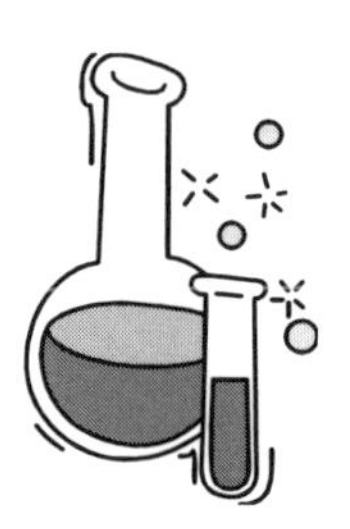

Materials:
- Tincture of iodine
- Various foods: bread, potato, macaroni, rice, crackers, flour, apple, orange, and milk

Safety Precautions: Iodine is toxic if ingested. Do not eat any foods that have been tested with iodine.

Procedure:
1. Place a drop of iodine on a piece of bread. Note the dark bluish-black color.
2. Test a variety of other foods in the same fashion.

Explanation: The iodine will turn a deep bluish-black color in the presence of starch. This is due to the formation of the starch-iodine complex. Many foods such as bread, rice, and pasta contain starch. Experiment with a variety of other foods. The formation of the dark blue color in the presence of starch is an example of a chemical change. Foods that lack starch, such as ripe fruit, will not test positive with iodine.

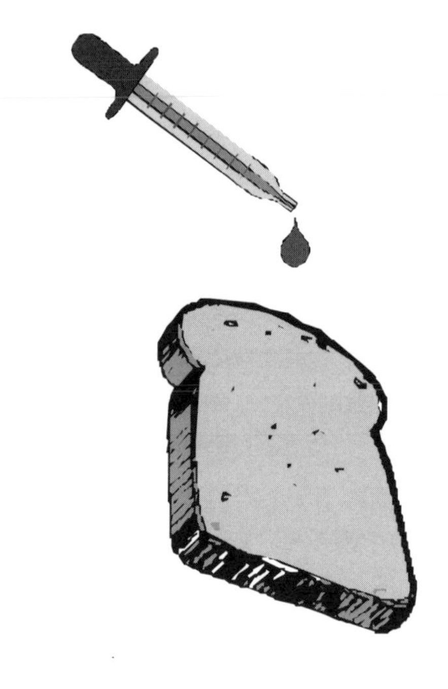

Chemical Reactions – Experiment # 25:

CONVERTING STARCH INTO SUGAR

Objective: To convert starch into sugar using saliva.

Materials:

- Tincture of iodine
- White paper
- Cotton swabs
- Crackers

Safety Precautions: Iodine is toxic if ingested.

Procedure:

1. Place several drops of iodine in the center of a piece of white paper.
2. Using a cotton swab, fashion the iodine into a large uniform circle on the paper.
3. Allow the iodine to dry. Note its color.
4. With a clean cotton swab, apply a liberal amount of saliva to the center of the circle of iodine on the paper.
5. Allow the saliva to dry. Contrast the color of the iodine where the saliva was applied to where it was not applied.
6. After washing your hands, put a cracker in your mouth, and keep it there without chewing for 5 minutes. How does it taste?

Explanation: The paper initially turns a deep blue-black color, because paper contains starch. The starch reacts with iodine to form the starch-iodine complex.

Saliva contains enzymes, which begin to digest food as it is chewed in the mouth. Enzymes are catalysts, which speed up the rate of a chemical reaction. One type of enzyme in saliva is known as amylase, which breaks down starch into simple sugars, such as maltose and glucose. These sugars do not react with iodine to form the starch-iodine complex. As a result, the color of the iodine on the paper will lighten considerably when saliva is applied. The enzymes in the saliva have converted the starch of the paper into sugar. This also explains why the cracker in your mouth tastes sweet after several minutes. The starch in the cracker is converted to sugar.

Chemical Reactions – Experiment # 26:
FORMING A GREEN PRECIPITATE

Objective: To produce a green solid by the addition of two liquids.

Materials:

- Steel wool
- Vinegar
- Household ammonia
- Glass jars

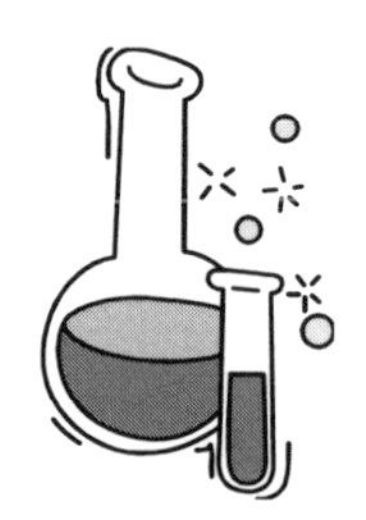

Safety Precautions: Ammonia is toxic if ingested. Do not inhale vapors of ammonia. Wear safety goggles.

Procedure:

1. Place a piece of steel wool in a glass jar.
2. Add enough vinegar to cover the steel wool.
3. Allow to remain undisturbed for one week. The substance that forms is iron(II) acetate.
4. Remove any solid steel wool from the liquid and dispose of in the trash.
5. Slowly add ammonia to this liquid until you note the formation of a thick green precipitate.

Explanation: The combination of iron and vinegar produces iron(II) acetate. The balanced chemical equation is as follows:

$$Fe_{(s)} + 2\,HC_2H_3O_{2(aq)} \Rightarrow Fe(CH_3COO)_{2(aq)} + H_{2(g)}$$

The addition of the ammonia (which contains OH^- ions in aqueous solution) to iron(II) acetate (which contains Fe^{2+} ions) results in the formation of a precipitate known as iron(II) hydroxide. The balanced chemical equation is as follows:

$$Fe^{2+}_{(aq)} + OH^-_{(aq)} \Rightarrow Fe(OH)_{2(s)}$$

Chemical Reactions – Experiment # 27:
DECOMPOSITION OF SUGAR

Objective: To liberate carbon from sugar.

Materials:

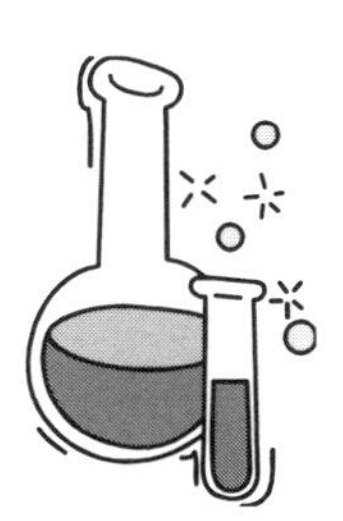

- Concentrated sulfuric acid (available in hardware stores as a drain cleaner – check the label)
- Sugar
- Sturdy glass jar

Safety Precautions: Adult supervision is required for this experiment Concentrated sulfuric acid is extremely corrosive to eyes and skin. Read all precautions on label before proceeding. Always wear safety goggles when dealing with acids. If any acid should come into contact with skin or eyes, flush with water and seek medical attention immediately. This experiment must be performed outdoors – toxic fumes are produced which are extremely irritating and will produce fits of coughing if inhaled. Observe from a distance of at least 5 feet away.

Procedure:

1. Place a half-cup of sugar in the glass jar.
2. Add an equal volume of sulfuric acid.
3. Quickly step back and remain at least 5 feet away.
4. After a minute or so, a solid black mass of carbon will rise up out of the jar several inches into the air.
5. After the mixture cools, wrap the container in several bags (wear gloves) and take to a hazardous waste facility. Be careful, since the black mass of carbon still contains some unreacted sulfuric acid.

Explanation: This is a fascinating chemical reaction to watch. The formula for table sugar (sucrose) is $C_{12}H_{22}O_{11}$. When the sugar reacts with the sulfuric acid, the hydrogen and oxygen are released as water vapor. As this water vapor condenses, its escape can be readily observed. It is the release of this gas that causes the carbon to rise – much like bread rises as it bakes due to the release of carbon dioxide gas produced by yeast. The little holes that are produced throughout the black mass of carbon are due to the escape of this water vapor. Note that the ratio of hydrogen to oxygen is the same as that found in H_2O. The balanced chemical equation is as follows:

$$C_{12}H_{22}O_{11(s)} \Rightarrow 12C_{(s)} + H_2O_{(g)}$$

In addition to water vapor, sulfur dioxide gas (SO_2) is also produced during this reaction, which is harmful to inhale. Since the hydrogen and oxygen are removed from the sugar, the only thing remaining is carbon, which is a solid, hard, black substance.

Do not touch the carbon, since it contains unreacted sulfuric acid. You will also note that if you touch the side of the glass jar, it is very hot, since this reaction is highly exothermic.

Chemical Reactions – Experiment # 28:
A RUBBERY EGG

Objective: To make a "rubbery" egg by removing its shell.

Materials:

- Egg
- Large plastic cup
- Vinegar

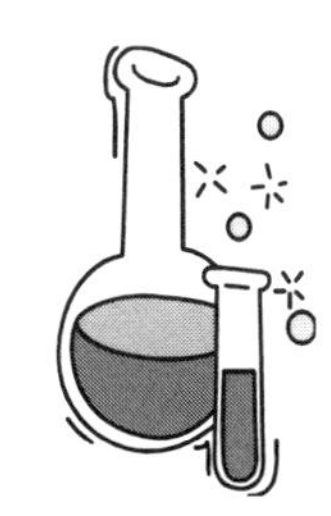

Safety Precautions: None

Procedure:

1. Place an egg in the cup.
2. Add enough vinegar to completely cover the egg.
3. Allow to remain undisturbed for several days.
4. Remove the egg and feel its texture.

Explanation: The vinegar, or acetic acid, reacts with the calcium carbonate in the eggshell – completely dissolving it. The initial bubbling from the egg is carbon dioxide gas being given off. The balanced chemical equation is as follows:

$$CaCO_{3(s)} + 2CH_3COOH_{(aq)} \Rightarrow Ca(CH_3COO)_{2(aq)} + H_2O_{(l)} + CO_{2(g)}$$

$CaCO_{3(s)}$	$2CH_3COOH_{(aq)}$	$Ca(CH_3COO)_{2(aq)}$	$H_2O_{(l)}$	$CO_{2(g)}$
calcium carbonate	acetic acid	calcium acetate	water	carbon dioxide

The end result is an egg without a shell. It can be bounced, but only very gently, or else it will break. You will also notice that the egg is somewhat larger than normal, as a result of fluid diffusing into the egg.

Chemical Reactions – Experiment # 29:

THE EFFECT OF BLEACH ON HAIR

Objective: To discover the effect of bleach on human hair.

Materials:

- Bleach
- Four plastic cups
- Human hair (available for free at any barber shop)
- Permanent marker

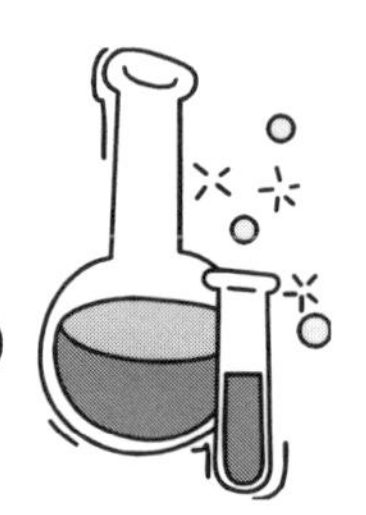

Safety Precautions: Bleach is very toxic if inhaled or ingested, or if it gets on the skin. Wear safety goggles. Be careful not to spill on clothing.

Procedure:

1. Label four cups 1-4. Fill cup #1 halfway with bleach. Label this cup as full strength.
2. In cup #2, add half water and half bleach. Label this cup as half bleach and half water.
3. In cup #3, add two parts water to one part bleach. Label this cup as two-thirds water and one-third bleach.
4. In cup #4, add three parts water to one part bleach. Label this cup as three-quarters water and one-quarter bleach.
5. Into each, place a generous amount of hair, making sure each hair sample is completely covered.
6. Make observations at hourly intervals throughout the day and make a final observation 24 hours later.

Explanation: The hair in the full strength bleach will most likely be completely dissolved within 24 hours. The lesser concentrations of bleach will have varying degrees of damage to the hair. Hair by itself is naturally acidic, so the bleach will chemically react with the hair in an attempt to neutralize it, since the bleach is a base. The bleach breaks down the bonds in the hair, softening the strands of hair, and eventually completely dissolves them.

Commercial hair removers, such as Nair, work the same way. If you read the label, you will find that they always contain a very strong base, which dissolves hair just like bleach.

Chemical Reactions – Experiment # 30:

THE EFFECT OF BLEACH ON FABRICS

Objective: To discover the effect of bleach on different types of fabrics.

Materials:

- Plastic cups
- Bleach
- Strips of different types of fabric, such as cotton, polyester, wool, or silk.

Safety Precautions: Bleach is very toxic if inhaled or ingested, or if it gets on the skin. Wear safety goggles. Be careful not to spill on clothing.

Procedure:

1. Pour equal amounts of bleach in separate cups. About a quarter-cup of bleach for each should be sufficient.
2. Into each cup place a strip of different fabric.
3. Observe the final results 24 hours later.

Explanation: The wool and silk will most likely be completely dissolved. The cotton and polyester, although bleached of color, probably will not be dissolved. The wool and silk are acidic, and the bleach, being basic, will neutralize these substances. In the process of neutralization, the fibers are weakened and eventually completely dissolved. The cotton and polyester, which are not acidic, are not affected as much. However, over time, these will eventually be broken down by the bleach. The pigments within these fabrics are removed, however, since the bleach chemically reacts with them and renders them colorless. Bleach is an excellent oxidizer, oxidizing the pigments into colorless compounds.

The same bleaching effect can be observed if a strip of fabric is placed over the mouth of a bottle of bleach and then the cap is replaced. Within a few hours, vapors of Cl_2 being released will bleach away the pigment from the fabric.

Chemical Reactions – Experiment # 31:
THE AMAZING SPARKING BALLS

Objective: To create an exothermic reaction by reacting aluminum and rust.

Materials:

- Two rusty iron balls about 2-3 inches in diameter (available from a scrapyard or junkyard)
- Aluminum foil

Safety Precautions: Wear safety goggles. Do this experiment away from flammable materials, since sparks are produced.

Procedure:

1. Wrap one of the rusty balls with a single layer of aluminum foil.
2. In a dark room, strike the foil-covered ball with the other rusty ball. Glancing blows in which the balls are struck together as they pass one another work better than direct blows.
3. Practice this technique until you are able to produce showers of sparks.

Explanation: This is a truly spectacular experiment that is worth the time it takes to find some rusty iron balls. The sparks are produced as a result of the reaction between the rust (Iron(III) oxide – Fe_2O_3) and the aluminum foil. Striking the two balls together produces sufficient frictional energy to initiate the reaction. The products are aluminum oxide (Al_2O_3), iron, and heat (which we observe as sparks). The balanced chemical reaction is as follows:

$$Fe_2O_{3(s)} + 2Al_{(s)} \Rightarrow Al_2O_{3(s)} + 2Fe_{(s)} + \text{heat}$$

This is an example of a thermite reaction, which is an extremely exothermic reaction between aluminum and certain metal oxides. Thermite reactions can generate temperatures up to 2200°C. This is hot enough to melt iron, which has a melting point of 1530°C. Thermite reactions have been used in welding, and to make fireworks, rockets, and bombs.

Chemical Reactions – Experiment # 32:
THE PRODUCTION OF ALCOHOL

Objective: To produce alcohol as a result of anaerobic fermentation.

Materials:

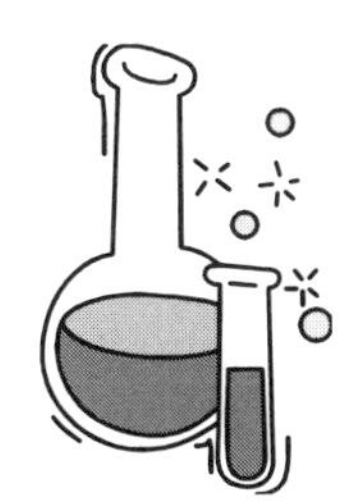

- 20 oz plastic soft drink bottle (do not use glass!)
- Dry, active yeast
- Molasses
- Measuring spoon
- Balloons

Safety Precautions: Perform this experiment only under adult supervision. The alcohol produced in this experiment is not fit for human consumption. Many additional waste products and harmful bacteria are produced, which can cause severe illness if consumed.

Procedure:

1. In the plastic bottle, place about 200 mL of warm water. The water should not be too hot, or it will kill the yeast. Hot water from the tap is sufficient.
2. Add three teaspoons of molasses and a half-teaspoon of yeast
3. Firmly attach the balloon over the mouth of the bottle.
4. Clearly label the bottle "POISON" and put it in a warm place where it will remain undisturbed.
5. Make daily observations for 14 days. Note the amount of gas released each day.
6. If the balloon breaks during the experiment, replace with another. If the balloon appears as though it may burst, remove to release the gas and then reattach.
7. After 14 days, remove the balloon and note the odor.

Explanation: The alcohol produced in this experiment is ethyl alcohol, or ethanol, which is produced by the process of anaerobic fermentation. Fermentation involves the breakdown of an organic compound into simpler substances due to the action of a catalyst. Anaerobic fermentation occurs in the absence of oxygen. This reaction is catalyzed by the enzyme zymase, which is produced by the yeast. The balanced chemical equation is as follows:

$$\underset{\text{glucose}}{C_6H_{12}O_{6(aq)}} \Rightarrow \underset{\text{ethanol}}{2C_2H_5OH_{(l)}} + 2CO_{2(g)}$$

The carbon dioxide causes the balloon to expand. Many other impurities are also produced during this experiment; the unpleasant odor is evidence of the large amount of bacteria produced.

Commercially, ethanol has a wide variety of uses. It is used in alcoholic beverages and thermometers. Gasoline often contains 10% ethanol. Ethanol is also used as a solvent in perfume, vanilla extract, iodine, and many other substances. An alcoholic solution is known as a tincture. If the solution is flammable, it is known as a spirit.

Repeat this experiment using other types of sugars, such as corn syrup or table sugar (sucrose). Also try flour or corn starch instead of molasses. What do you think will happen? (Hint: Ethanol is also known as grain alcohol.)

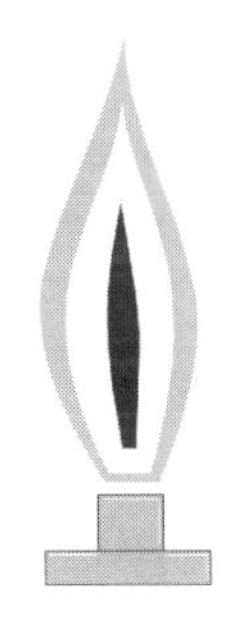

CHAPTER 6
COMBUSTION

Combustion is burning in the presence of oxygen. Fires require three elements in order to burn: heat, fuel, and oxygen. Without any of the above, fires will cease to burn.

There are four classes of fires: A, B, C, and D. If you look on the side of any fire extinguisher, it will tell you what type of fire it is labeled for.

Class A fires are the most common. They are formed when paper or wood burns. Water, carbon dioxide, and dry powders can be used to extinguish a class A fire.

Class B fires are formed when a flammable liquid, such as oil or grease, burns. Never put water on a grease fire! It will only serve to spread it more. Carbon dioxide and dry powder can be used on a class B fire.

Class C fires are electrical fires. Never put water on an electrical fire! Carbon dioxide or dry powder can be used to extinguish a class C fire.

Class D fires occur when combustible metals burn. Only dry powder should be used on this type of fire. Some metals, such as magnesium, burn very well in carbon dioxide. Adding water to a sodium fire would only make it worse.

We will safely explore the awesome nature of combustion in this chapter . . .

Combustion – Experiment # 1:
PRODUCTS OF COMBUSTION

Objective: To observe the products of the combustion of a candle.

Materials:
- Large candle
- Matches
- Glass jar

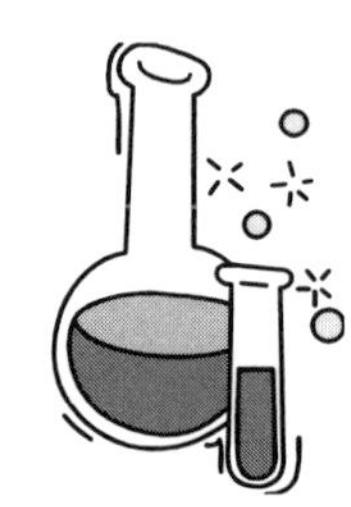

Safety Precautions: Always exercise caution when using open flames.

Procedure:
1. Hold the bottom of a glass jar just over the tip of a candle flame. You will notice that a thin layer of condensation immediately forms.
2. Continue to hold the jar over the flame for about 10 seconds. A thick black coating will form on the jar.

Explanation: A typical candle is composed of paraffin or wax – which is a hydrocarbon. A hydrocarbon is a compound composed only of hydrogen and carbon. A typical paraffin molecule will have a formula of $C_{25}H_{52}$. When a candle burns, it reacts with oxygen in the air to yield carbon dioxide and water vapor. The balanced chemical equation is as follows:

$$C_{25}H_{52(s)} \quad + \quad 38O_{2(g)} \Rightarrow 25CO_{2(g)} \quad + \quad 26\,H_2O_{(g)}$$

The first substance formed on the jar is a thin layer of water droplets that form during the combustion of the candle. Upon hitting the cool jar, this gaseous water condenses. You may have noticed condensation dripping from the tailpipe of a car. This water is formed due to the combustion of gasoline. Gasoline is also a hydrocarbon.

As the candle is allowed to remain in contact with the glass, the glass becomes covered with a thick black layer of what is commonly known as "soot." Soot is unburned carbon particles. Soot will cause a candle flame to burn a yellow color. Gas stoves always burn with a blue flame – which is much cleaner and hotter than a yellow flame. A blue flame will seldom leave a soot deposit on your pots and pans, since nearly all of the carbon is consumed in the flame.

Combustion – Experiment # 2:
THE MYSTERIOUS RISING WATER

Objective: To demonstrate that combustion requires oxygen.

Materials:

- Candles
- Matches
- Shallow metal pie pan
- Metal jar lid
- Transparent drinking glass

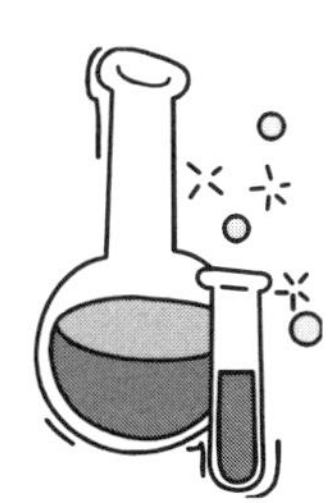

Safety Precautions: Always exercise caution when using matches and open flames.

Procedure:

1. Firmly attach a candle to the metal jar lid by melting some wax on the lid and placing another candle in the wax.
2. Fill a pie pan with water and place the jar lid with the attached candle in the center of the pan. Make sure the water level only rises about halfway up the candle.
3. Light the candle, and then place the glass over the candle, allowing the mouth of the glass to rest on the bottom of the pie pan.
4. The candle will burn for a few seconds, and as soon as it goes out, the water level in the glass will rise dramatically!

Explanation: The rise in water is sudden, quick, and quite surprising at first sight. As the air inside the glass is heated, its molecules move faster, causing the air to expand. This expanding air has nowhere to go but downward and into the water. Careful observation will reveal bubbles of air escaping as the glass is put over the flame. This creates a region of reduced pressure inside the glass. Since air pressure is now greater outside of the glass than inside, air pressure pushing down on the water forces the water into the glass.

This experiment is sometimes erroneously used to demonstrate that as the candle burns, oxygen is consumed, since the candle shortly goes out after the glass is put over it. Since the atmosphere is one-fifth oxygen, some argue that this explains why the water level in the glass rises to fill approximately one-fifth of the volume of the glass.

However, this explanation is not sufficient. As oxygen is consumed by the flame other gases are produced in their place, namely water and carbon dioxide. These other gases will occupy approximately the same volume as the oxygen. Therefore the reduction in pressure within the glass cannot be explained by the consumption of oxygen by the fire, but only due to the removal of air by heating.

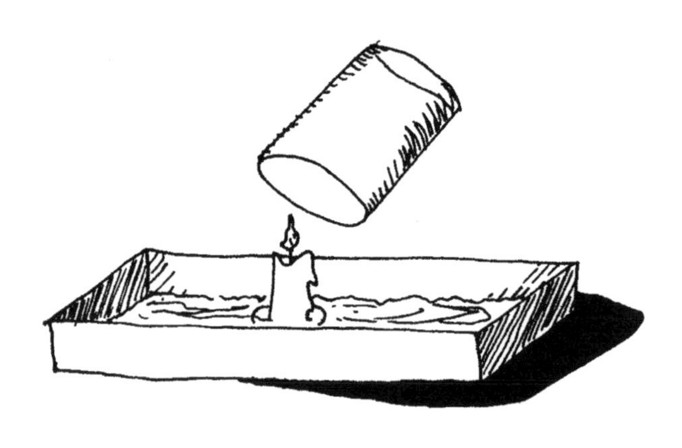

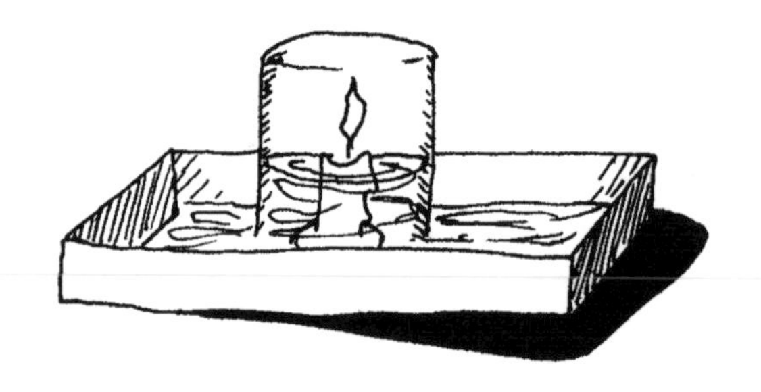

Combustion – Experiment # 3:

FLAMING POWDER

Objective: To demonstrate that increasing the surface area of a substance greatly increases its rate of combustion.

Materials:

- Candle
- Matches
- Flour or Lycopodium powder (Better results can be obtained using Lycopodium powder, which is available in magician and novelty shops, sometimes under the name of "Dragon's Breath." However, flour will work fine, as long as it is dry.)
- Funnel
- Plastic aquarium tubing (3 foot length)

Safety Precautions: Make sure there are no flammable or combustible materials nearby! A huge column of flames may be produced. It is best to practice this experiment outside so that you fully understand the ramifications, before attempting indoors. Wear safety goggles. Have a fire extinguisher nearby.

Procedure:

1. Make sure the funnel fits tightly into the end of the plastic tubing.
2. Place about a teaspoon of flour or quarter-teaspoon of Lycopodium powder into the funnel. You will need to experiment with the amounts to produce the desired results. Begin with a small amount.
3. Place the funnel near the candle flame, and exhale vigorously. A large burst of intense flames should be produced.

Explanation: This experiment will produce spectacular effects, which are especially visible in a dark room. If flour does not work adequately, it is well worth a search for some Lycopodium powder, which always produces a tremendous burst of flames. This experiment is successful, not due to any particular combustible properties of flour or Lycopodium, but instead due to the very fineness of the powders and the tremendous amount of surface area that comes into contact with the flame. Combustion is a chemical reaction, and the more surface area that is exposed to the flame, the faster the rate of reaction. Lycopodium, which are the spores of Club Moss, work better than flour because its particles are much finer. Explosions have been known to occur in grain elevators where a great deal of fine dust is generated. Many metals in powder form are very combustible if they come into contact with a flame.

Experiment with other fine powders, such as powdered sugar, cinnamon, and non-dairy coffee creamer. But please be careful!

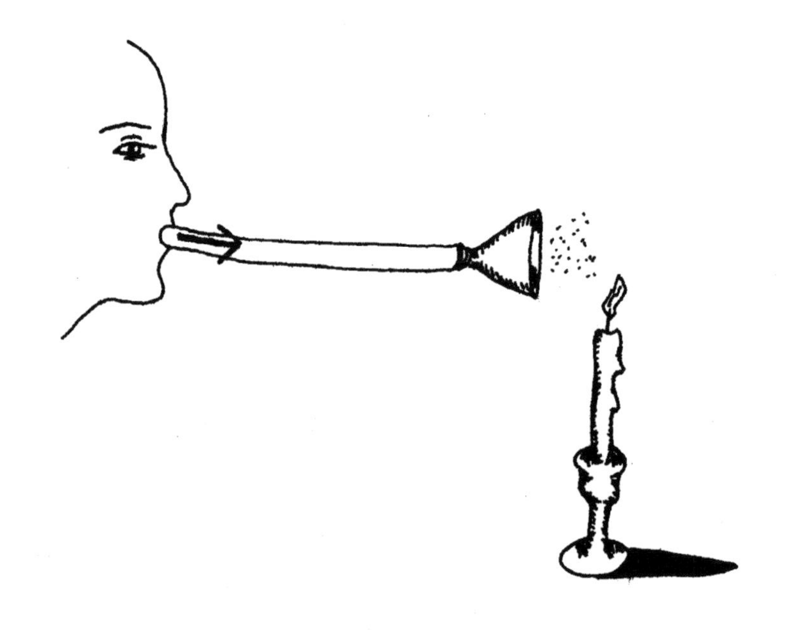

Combustion – Experiment # 4:

THE EXPLODING CAN

Objective: To simulate a grain elevator explosion.

Materials:

- Flour or Lycopodium powder (see previous experiment)
- Large coffee can with lid
- One inch wide by one inch high candle
- Drill
- Funnel
- 3 feet of plastic aquarium tubing
- Matches

Safety Precautions: Wear safety goggles. Stand back as far as possible. Remove all flammable or combustible substances from the area. Have a fire extinguisher nearby.

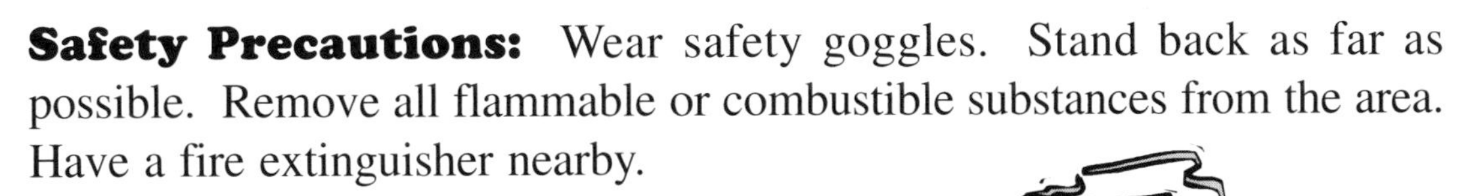

Procedure:

1. Drill a hole in the coffee can about 1 inch from the bottom, so that the plastic tubing fits through snugly.
2. Inside the can, attach the funnel to the rubber tubing, and allow the funnel to rest on the bottom of the can.
3. Place about a teaspoon of flour or quarter-teaspoon of Lycopodium in the funnel. You will need to experiment with the amounts to achieve the desired results.
4. Place the candle in the bottom of the can, so that the mouth of the funnel is directly facing the candle.
5. Light the candle, quickly fasten the lid, and then blow into the tubing. A burst of flames should be produced inside the can, which will send the lid flying several feet into the air. If this experiment will not work with flour, try Lycopodium. It may take some practice, but the end result is well worth the effect.
6. If the flame burns a hole in the lid, repair the lid with duct tape.

Explanation: The same principle is at work here as in the previous experiment. Any very fine powder is extremely combustible due to its large surface area. The flames produced when the powder is blown on the candle quickly heat the air in the can. This causes the air within the can to rapidly expand. This expansion creates a tremendous build-up of pressure within the can, which is sufficient to send the lid flying into the air. The combustion of the powder itself also produces gaseous products, which further contribute to the increase of pressure within the can.

Explosions like this sometimes occur on a much grander scale in grain elevators or other confined areas where a great deal of dust is generated and accidentally ignited.

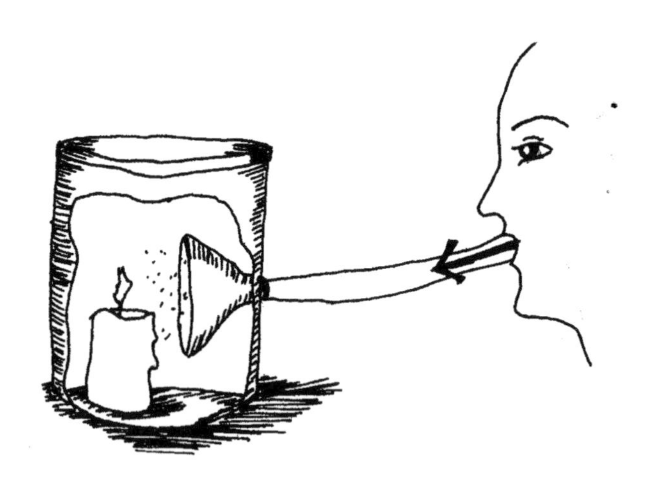

Combustion – Experiment # 5:
BURNING WATER

Objective: To demonstrate that liquids can burn while floating in water.

Materials:

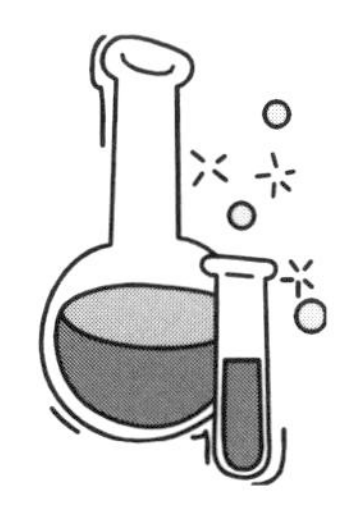

- Old metal pan with lid
- Lighter fluid
- Matches

Safety Precautions: Lighter fluid is very flammable – only light under controlled conditions. Never pour lighter fluid directly onto an open flame! Always wear safety goggles when dealing with fire. Keep a fire extinguisher nearby.

Procedure:
1. Squirt some lighter fluid into the bottom of the pan.
2. Fill the pan halfway with water.
3. Throw a lit match into the pan. It should catch on fire.
4. Smother the fire by covering with the lid.

Explanation: This experiment demonstrates several principles. Lighter fluid does not mix with water, and since it is less dense than water, it floats. However, as long as oxygen is present, lighter fluid will still burn while floating on water. By covering the pan, the oxygen supply is depleted, so the fire goes out.

You should never put water on a Class B fire (grease or other flammable liquids). Since water and grease do not mix, the water will simply spread the fire. If possible, extinguish a grease fire by smothering it.

In 1969 in Cleveland, Ohio, the Cuyahoga River actually caught on fire when an oil slick was accidentally ignited, perhaps by sparks from a train passing overhead on a bridge. The column of flames produced reached the height of a five story building and burned two bridges. This fire was only possible because oil is less dense than water.

Combustion – Experiment # 6:

BURNING A MARSHMALLOW

Objective: To discover the products arising from the combustion of a marshmallow.

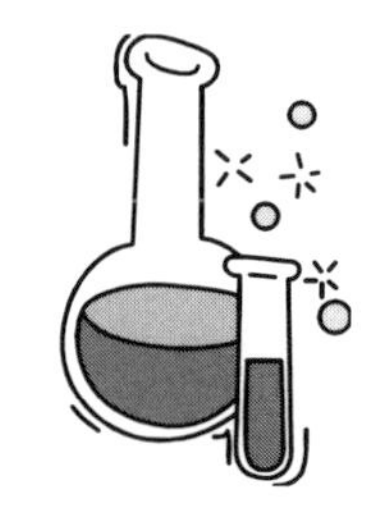

Materials:

- Marshmallows
- Sharpened stick or wooden bamboo skewer
- Propane torch or campfire

Safety Precautions: Always exercise caution around open flames.

Procedure:

1. Carefully insert a sharpened stick through a marshmallow.
2. Hold over the flame until it has completely burned – there should be no "white stuff" remaining at all.
3. After it cools, crumble the black marshmallow with your hand. It will be hollow.

Explanation: Marshmallows are composed primarily of sucrose, or cane sugar. The formula for sucrose is $C_{12}H_{22}O_{11}$. The hydrogen and oxygen ratio in sugar is the same as that for water or H_2O – a 2:1 ratio. As the marshmallow burns, its hydrogen and oxygen is released as water vapor. The black part of the marshmallow represents the unburned carbon in the marshmallow – which is all that remains once the hydrogen and oxygen leave. The release of these gases causes the marshmallow to expand considerably

The mass of the remaining black part of the marshmallow is about 40% of the original mass of the marshmallow. The rest of the mass has been released as water vapor.

Anytime food is burned and it turns black, the carbon is being separated from the rest of the food. If something is burned long enough, nearly all that remains is carbon – definitely not very tasty!

CHAPTER 7
ACIDS AND BASES

Whether we realize it or not, acids and bases play a major role in our lives. Stomach acid aids in the digestion of food. Sulfuric acid is necessary for the proper operation of an automobile battery. The orange juice we drink for breakfast and the soda we later consume all contain acid. Milk turns sour due to lactic acid. Acids have a pH less than 7 at 25°C.

Soap, detergent, ammonia, bleach, and nearly everything we clean with are examples of bases. Bases taste bitter and tend to feel slippery. Bases have a pH greater than 7 at 25°C.

Chemically speaking, acids produce hydrogen ions (H^+) in aqueous solution. Bases produce hydroxide ions (OH^-) in aqueous solution. Acids can also be thought of as proton donors, and bases as proton acceptors.

There are a wide range of fascinating experiments involving acids and bases. A few are presented in this chapter . . .

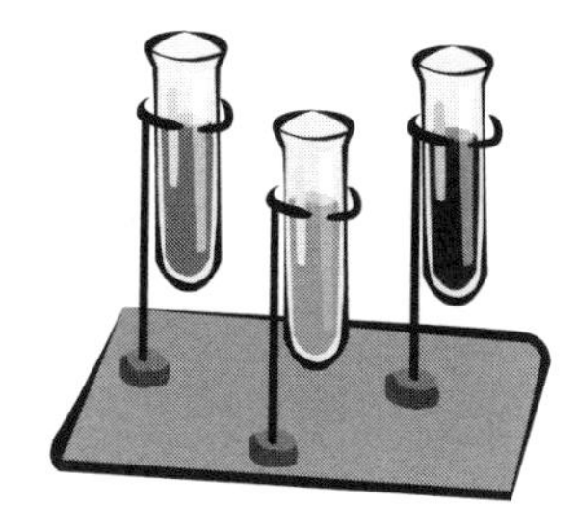

Acids and Bases – Experiment # 1:

RED CABBAGE AS AN INDICATOR

Objective: To determine if a substance is an acid or a base.

Materials:

- Head of red cabbage
- Pan
- Stove or hotplate
- Several transparent plastic cups
- Vinegar
- Household ammonia
- Isopropyl rubbing alcohol
- Various household substances

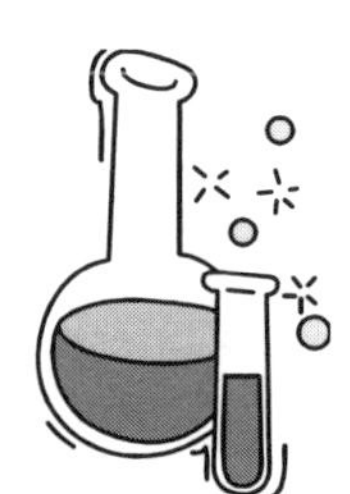

Safety Precautions: Exercise caution when using the stove. Be careful when boiling liquids. Many household cleaning substances are harmful if swallowed, inhaled, or spilled on skin. Always wear safety goggles when handling any chemical substances. Read the labels of all household substances before using.

Procedure:

1. Tear off a few leaves from a head of red cabbage and place them in a pan. Add enough water to cover. Heat to boiling and allow the water to boil until most of the pigment has been removed from the leaves. More water can be added as needed.
2. Pour the red cabbage juice into several transparent plastic cups. The juice can be diluted quite a bit with water and still maintain its deep purple to blue color.
3. Begin by pouring some red cabbage juice into 3 separate containers.
4. Into the first container, pour some vinegar. You will notice the color change to red.
5. To the second container, add some ammonia. You will notice the color change to green.
6. Into the third container, add some alcohol. You should notice no color change at all.
7. Experiment with as many other liquid substances as you can find.

8. Test some solid substances as well, such as baking soda, laundry detergent, salt, and sugar.

Explanation: Red cabbage is an excellent example of an acid-base indicator. Due to the anthocyanin pigment in red cabbage, it changes colors in the presence of an acid or a base. Vinegar, or acetic acid, turns the cabbage juice red. Any other acid will have the same effect. Some other acidic substances to test are orange juice (citric acid), apple juice (malic acid), and soft drinks (carbonic and phosphoric acid).

The ammonia, being a base, turns the cabbage juice green. Other examples of basic or alkaline substances are soaps, detergents, bleach, and other cleaning agents.

Alcohol, being neutral, has no effect on the cabbage juice. Can you discover other neutral substances?

Related Experiments: There are many other natural acid-base indicators. Blueberries, grape skins, beets, and many other deeply colored fruits or vegetables will generally serve as an acid-base indicator. Use the same procedure to extract the juice.

Acids and Bases – Experiment # 2:

NEUTRALIZATION

Objective: To neutralize an acid by adding a base, and vice versa.

Materials:

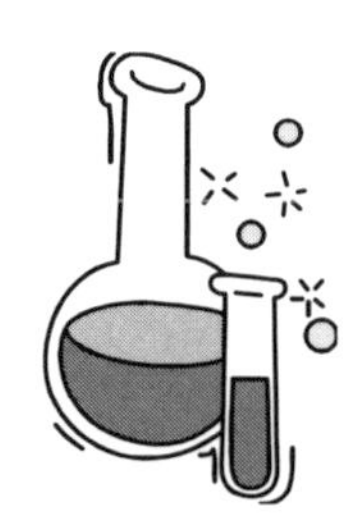

- Red cabbage juice (see previous experiment)
- Vinegar
- Household ammonia
- Several transparent plastic cups
- Eyedropper

Safety Precautions: Exercise caution when using ammonia or any other household substance. Always read the label before using. Wear safety goggles.

Procedure:

1. Add some red cabbage juice to a cup. Note its color.
2. Add ammonia dropwise until the color changes to green. The substance is now basic, or alkaline.
3. Now add vinegar dropwise until it reverts back to its original purplish color. The substance is now neutral again.
4. Repeat the experiment, using vinegar to make the cabbage juice acidic, and then adding ammonia to neutralize it.

Explanation: An acid will neutralize a base, and vice-versa. One definition of an acid is a substance that donates a hydrogen ion (H^+) in water. In other words, when an acid is added to water, hydrogen ions are given off. When water is added to a base, hydroxide ions (OH^-) are released. When a hydrogen ion joins with a hydroxide ion, the end result is water, a neutral substance. The reaction is as follows:

$$2H^+_{(aq)} + OH^-_{(aq)} \Rightarrow H_2O_{(l)}$$

Acids and Bases – Experiment # 3:

ACID BREATH

Objective: To create an acid by blowing into a glass of water.

Materials:

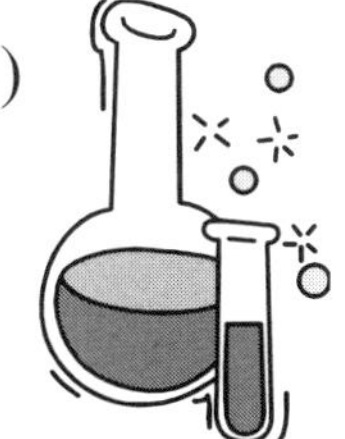

- Red cabbage juice (see Acids and Bases – Experiment # 1)
- Drinking straw
- Two transparent plastic cups

Safety Precautions: None

Procedure:

1. Add an equal amount (about 30 mL) of red cabbage juice to two drinking glasses.
2. Allow one glass to remain undisturbed.
3. Using the straw, exhale (do not inhale!) into the glass of cabbage juice for several minutes. You will slowly begin to see a color change. Compare the color in this glass to the glass that was left undisturbed.

Explanation: When you exhale into the water, you are exhaling carbon dioxide (CO_2) gas. CO_2 reacts with water to form carbonic acid. The reaction is as follows:

$$CO_{2(g)} + H_2O_{(l)} \Rightarrow H_2CO_{3(aq)}$$

The formation of carbonic acid changes the neutral cabbage solution to one that is slightly acidic. Thus the original blue-purple color will gradually turn to red as carbonic acid is formed.

Bromothymol blue indicator solution can also be used in place of the red cabbage juice. It is generally available from pool stores or pet shops that carry aquarium supplies. It is blue in a neutral or basic solution, and turns yellow in the presence of an acid.

Acids and Bases – Experiment # 4:

AMAZING ALKA-SELTZER REACTION

Objective: To observe the neutralization of a basic solution by adding Alka-Seltzer.

Materials:

- Red cabbage juice (see Acids and Bases – Experiment # 1)
- Transparent plastic cup
- Alka-Seltzer tablets
- Household ammonia

Safety Precautions: Ammonia is toxic if ingested. Do not inhale vapors. Wear safety goggles.

Procedure:

1. Fill a plastic cup nearly full with red cabbage juice.
2. Add just enough ammonia dropwise to make the solution basic (a greenish color).
3. Drop in the Alka-Seltzer tablet; you should observe the color change to purple and finally to red. Drop in more tablets if you do not see the full range of color changes.

Explanation: As the Alka-Seltzer tablet dissolves in water, carbon dioxide gas is released. This CO_2 gas combines with water to form carbonic acid (see previous experiment). The formation of carbonic acid neutralizes the basic solution and finally causes it to become acidic, thus the color of the red cabbage juice changes from green to purple to red. The neutralization of ammonia with carbonic acid can be represented by the following balanced chemical equation:

$$H_2CO_{3(aq)} + 2NH_4{}^+{}_{(aq)} + 2OH^-{}_{(aq)} \Rightarrow (NH_4)_2CO_{3(aq)} + 2H_2O_{(l)}$$

Repeat this experiment with other acid-base indicators. Bromothymol blue works especially well (see previous experiment).

Acids and Bases – Experiment # 5:

WHICH ANTACID IS MOST EFFECTIVE?

Objective: To determine which brand of commercial antacid is the most effective.

Materials:

- Red cabbage juice (see Acids and Bases – Experiment # 1)
- Several transparent plastic cups
- Eyedropper
- Several brands of commercial antacids:
    - Tums
    - Rolaids
    - Maalox
    - Generic brand
    - Others

Safety Precautions: None

Procedure:

1. Pour equal amounts (about 30 mL) of red cabbage juice into transparent cups.
2. Add just enough vinegar dropwise to each cup until each turns red. Be sure to add the same amount of vinegar to each cup.
3. Allow one cup (with vinegar added) to remain undisturbed, to serve as a control.
4. In the remaining cups, place a crushed antacid tablet in each.
5. Observe each closely and note how long it takes for the different antacids to cause a color change. Compare each sample to the control.

Explanation: Commercial antacids are designed to neutralize excess stomach acid. They will generally contain a hydroxide and/or a carbonate, each effective at neutralizing acid. This experiment can determine several things: Does the antacid actually work? How fast does it work? How much acid does it neutralize?

Careful observations of the above experiment can answer each of these questions. As the acid is neutralized, the color of the cabbage juice will change from red to blue.

Acids and Bases – Experiment # 6:

NEUTRALIZATION WITH BAKING SODA

Objective: To discover the neutralizing effects of baking soda.

Materials:

- Red cabbage juice (see Acids and Bases – Experiment # 1)
- Transparent cup
- Baking soda
- Vinegar
- Teaspoon

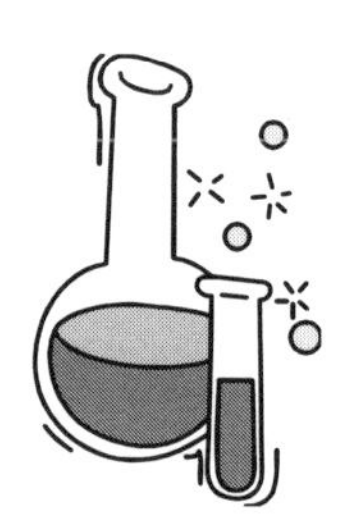

Safety Precautions: None

Procedure:

1. Pour some red cabbage juice into a cup.
2. Add vinegar until it turns a red color.
3. Add baking soda to the vinegar until the cabbage juice turns back to its neutral color (blue).

Explanation: The vinegar, being acidic, turns the cabbage juice red. The baking soda, upon reacting with the vinegar, changes the vinegar into a neutral substance. This experiment demonstrates that a base, such as baking soda (sodium bicarbonate), is effective at neutralizing an acid, such as vinegar. The fizzing that you observed during this reaction was the CO_2 gas that was being given off. The balanced chemical equation is as follows:

$$\underset{\text{baking soda}}{NaHCO_{3(s)}} + \underset{\text{vinegar}}{HC_2H_3O_{2(aq)}} \Rightarrow \underset{\text{sodium acetate}}{NaCH_3COO_{(aq)}} + \underset{\text{water}}{H_2O_{(l)}} + \underset{\text{carbon dioxide}}{CO_{2(g)}}$$

Acids and Bases – Experiment # 7:

MAKE YOUR OWN BASE INDICATOR

Objective: To make a highly sensitive and very effective basic indicator.

Materials:

- Turmeric (available from the spice section of the grocery store)
- 70% isopropyl alcohol
- Plastic cups
- Household ammonia
- Vinegar
- Alcohol
- Various household cleaning solutions

Safety Precautions: Many household cleaning substances are harmful if swallowed, inhaled, or spilled on skin. Always wear safety goggles when handling any chemical substances. Read the labels of all household substances before using.

Procedure:

1. Add a teaspoon of turmeric to about 50 mL of isopropyl alcohol.
2. Pour a small amount of ammonia in a plastic cup. Add a few drops of the turmeric solution. A red color will appear.
3. Repeat the above test with vinegar (an acid) and alcohol (neutral). What do you observe?
4. Add some vinegar to the red ammonia solution. What do you observe?

Explanation: Turmeric is derived from the root of a plant grown in the East Indies, and is also used to dye mustard yellow. Turmeric is an excellent basic or alkaline indicator. It turns red in the presence of a base, such as ammonia. An acidic or neutral solution will not change colors in turmeric. Adding an acid such as vinegar will cause the red solution to turn colorless because the basic ammonia is neutralized.

Acids and Bases – Experiment # 8:

MAGIC GOLDENROD PAPER

Objective: To discover the effect of a basic substance on goldenrod paper.

Materials:

- Several sheets of goldenrod paper (available at an office supply store – not all brands work; test with a basic substance before buying)
- Vinegar
- Household ammonia
- Eyedropper

Safety Precautions: Ammonia is toxic if ingested or the fumes inhaled. Wear safety goggles.

Procedure:

1. Drop several drops of vinegar on the goldenrod paper. Observe.
2. Drop several drops of ammonia on the goldenrod paper. Observe.
3. Now place several drops of vinegar on top of where the ammonia was placed on the paper. Observe.

Explanation: Goldenrod paper is an excellent base indicator – turning bright red in the presence of a basic substance. This is due to the use of turmeric to dye the paper yellow (see previous experiment). Acids have no effect on this paper, but they will cause the paper to turn yellow again if placed on top of the ammonia. Since acids neutralize bases, the red color from the ammonia will disappear when the ammonia is neutralized.

Related Experiments: There are numerous variations on the above experiment:

1. You can dip your finger in ammonia and pretend to write with a "bloody" finger.
2. You can tape a message in scotch tape, and spray the paper with ammonia (use a spray bottle) to reveal the message. The message under the tape will not change color.
3. Use a small paintbrush to "paint" a message on the goldenrod paper with ammonia.

4. Place the paper over the mouth of a bottle of ammonia. The fumes from the ammonia immediately turn the paper red.
5. If sodium silicate solution (available from the hardware store as water glass, in the paint section) is used to write on the paper, it will create a permanent red glossy image on the goldenrod paper.

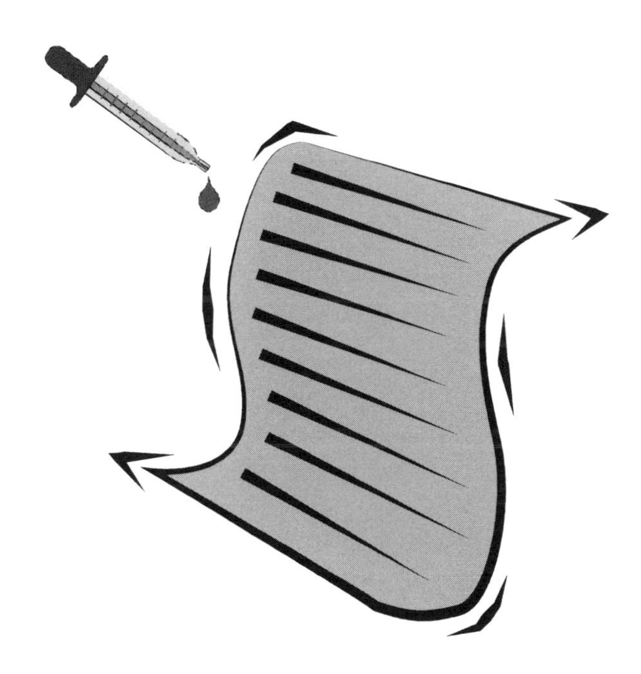

Acids and Bases – Experiment # 9:

MAGIC WRITING

Objective: To use phenolphthalein solution as "invisible ink," which can be revealed by applying a base.

Materials:

- Phenolphthalein solution (available from a chemical supply house – formerly available in Ex-Lax laxative tablets)
- Q-tips or small paintbrush
- White paper
- Spray bottle of ammonia
- Spray bottle of vinegar

Safety Precautions: Do not inhale fumes of ammonia; they can be extremely irritating. Do this experiment outside or in a well-ventilated area. Wear safety goggles.

Procedure:

1. Use the Q-tip or paintbrush to write a message on the piece of white paper.
2. Allow to dry for a few minutes. It should become completely invisible.
3. Spray with ammonia to reveal your message.
4. Spray with vinegar to conceal the message again.

Explanation: Since the phenolphthalein is a base indicator, the ammonia reveals the message. Since vinegar is an acid, it will neutralize the ammonia and cause the message to become invisible.

(Author's note: Phenolphthalein is not readily available to the public, since it is no longer used as the active ingredient in Ex-Lax. It has been removed from laxative tablets, due to the recent classification of phenolphthalein as a reasonably anticipated human carcinogen. A search on the Internet should reveal a readily available source of phenolphthalein at an inexpensive price. This experiment has been retained in the revised version of this book, as Experiment 151, since it is a favorite of many.)

Acids and Bases – Experiment # 10:

pH TESTING

Objective: To determine how acidic or basic a substance is by recording its pH.

Materials:

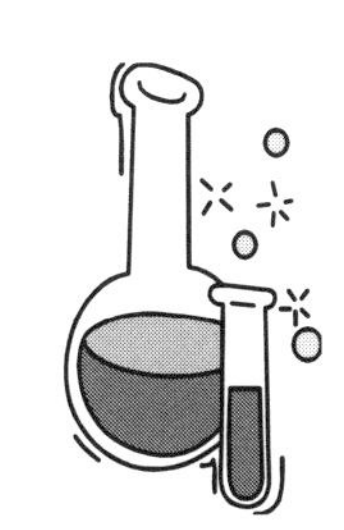

- pH paper (available from a pool supply store or pet shop that carries aquarium supplies)
- Transparent plastic cups
- Various liquid household substances

Safety Precautions: Always read labels of all household substances before using. Wear safety goggles.

Procedure:

1. Obtain as many liquid household substances as you can, pouring a small amount of each into a plastic cup.
2. Test each with a piece of pH paper, by comparing the color of the paper with the color code on the vial.
3. Arrange your substances in order from most acidic to most basic.

Explanation: The pH scale is a convenient way of measuring how acidic or basic a substance is. It is based on the number of hydrogen ions found in a given solution. A low pH represents a highly acidic substance. A high pH represents a more alkaline or basic substance. A pH of 7 is neutral at 25°C. A pH below 7 at 25°C is acidic. A pH greater than 7 at 25°C is alkaline. Battery acid (H_2SO_4) will have a pH of 1, and lye (NaOH) will have a pH of 13. Pure water is neutral, with a pH of 7.

The pH scale is a logarithmic scale – this means that a substance with a pH of 4 is 10 times more acidic than one with a pH of 5. A substance with a pH of 3 is 100 times more acidic than one with a pH of 5.

Related Experiments: Use the pH paper to record the pH of the substances you tested in experiment # 1 with the red cabbage indicator. You can now know the approximate pH range of a substance by seeing what color it turns in red cabbage juice.

Acids and Bases – Experiment # 11:

IS YOUR RAINFALL ACIDIC?

Objective: To determine if rainfall is acidic.

Materials:

- Empty plastic or glass containers
- pH paper
- Water samples from various sources

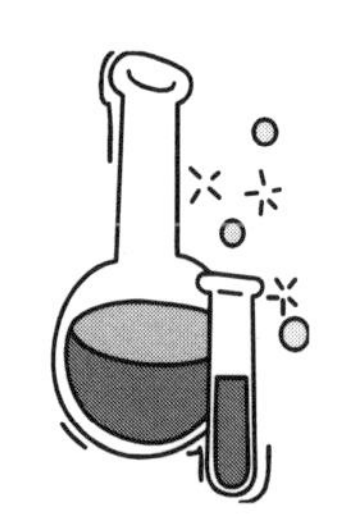

Safety Precautions: None

Procedure:

1. Obtain rainwater samples by placing a glass or plastic container outdoors and collecting rainwater.
2. Test the pH with the pH paper.
3. Continue testing rainwater over a period of several weeks to see if the pH changes.
4. In the winter, record the pH of melted snow.
5. Test other water samples: rivers, lakes, ponds, or tap water. Does their pH vary?

Explanation: Many industrialized areas of the world have very acidic rainfall due to pollutants from cars and factories. Even natural rainfall will be slightly acidic, due to the rain filtering out carbon dioxide from the atmosphere and forming carbonic acid. Natural rainfall will generally have a pH ranging from 5.0 to 5.6. How does your rainfall compare? There are a number of factors that can affect the pH of a body of water – type of soil, rocks, or vegetation. Pollution may also play a big factor.

If the pH of rainfall is below 5.0, it is considered acid rain. Acid rain forms when pollutants in the air combine with rainfall to form dilute solutions of acid. The two most common forms of acid rain contain either sulfuric acid (H_2SO_4) or nitric acid (HNO_3).

Sulfuric acid forms when sulfur dioxide (SO_2) combines with oxygen to form sulfur trioxide (SO_3). Sulfur trioxide then combines with water to form sulfuric acid. The most common source of SO_2 is coal-burning power plants. The balanced chemical equations are as follows:

$$2SO_{2(g)} + O_{2(g)} \Rightarrow 2SO_{3(g)}$$

$$SO_{3(g)} + H_2O_{(l)} \Rightarrow H_2SO_{4(aq)}$$

Nitric acid is formed when nitrogen oxides, such as NO_2, combine with water. The most common source of nitrogen oxide pollution is automobile exhaust. A common component of auto exhaust is nitric oxide (NO), which combines with O_2 to form nitrogen dioxide (NO_2). The NO_2 then combines with water to form a mixture of nitric acid (HNO_3) and nitrous acid (HNO_2). The balanced chemical equations are as follows:

$$2NO_{(g)} + O_{2(g)} \Rightarrow 2NO_{2(g)}$$

$$2NO_{2(g)} + H_2O_{(l)} \Rightarrow HNO_{3(aq)} + HNO_{2(aq)}$$

Acid rain has had a corrosive effect on many monuments and statues, especially those made of marble. Fish and other aquatic life will begin to die if the pH goes below 6.5. Few aquatic species can survive below a pH of 5.0. The lowest pH rainfall ever recorded was in 1982 over the Great Lakes, where a pH of 2.83 was recorded. This is the approximate pH of lemon juice!

Acids and Bases – Experiment # 12:

TESTING YOUR SOIL pH

Objective: To determine the pH of different soil types, and the factors that lead to this pH.

Materials:

- pH paper
- Plastic bags
- Permanent marker
- Plastic cups
- Distilled water

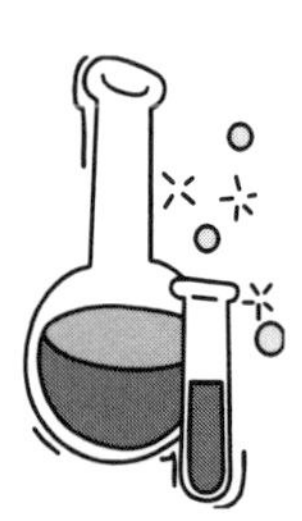

Safety Precautions: None

Procedure:

1. Collect soil samples from as many sources as possible: lawn, meadow, coniferous forest, deciduous forest, etc. Place each sample in a plastic bag and label.
2. Place a small amount of each soil sample in a plastic cup, and add an equal amount of distilled water to each. Mix thoroughly.
3. Record the pH of each soil sample.

Explanation: Soil type is generally determined by the rocks below the soil and the vegetation that grows above the soil. Certain types of rocks, such as limestone, are highly alkaline, and will contribute to a high pH. Other types of soils, such as the red clay of the southern states, tend to contain large amounts of iron oxide, and are more acidic.

The needles of coniferous trees fall to the ground and decay, tending to make the soil more acidic. The leaves of deciduous trees tend to be more neutral as they decay. The pH of your lawn may depend upon what types of fertilizers have been added to the lawn, or the underlying soil type.

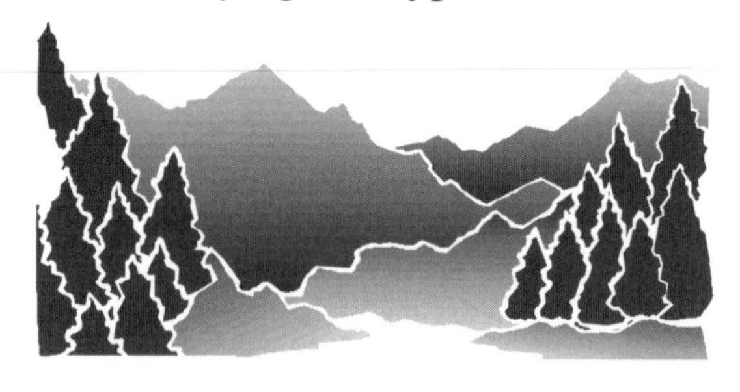

Acids and Bases – Experiment # 13:

CHANGING YOUR SOIL pH

Objective: To raise or lower the pH of your soil by the addition of fertilizers.

Materials:

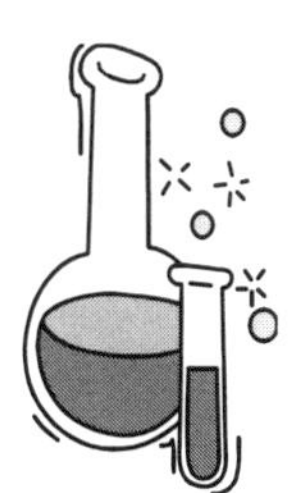

- pH paper
- Plastic cups
- Soil sample from lawn
- Distilled water
- Variety of fertilizers (from garden center):
  - Aluminum Sulfate
  - Lime
  - Miracle-Gro (or any other lawn or tree fertilizer)

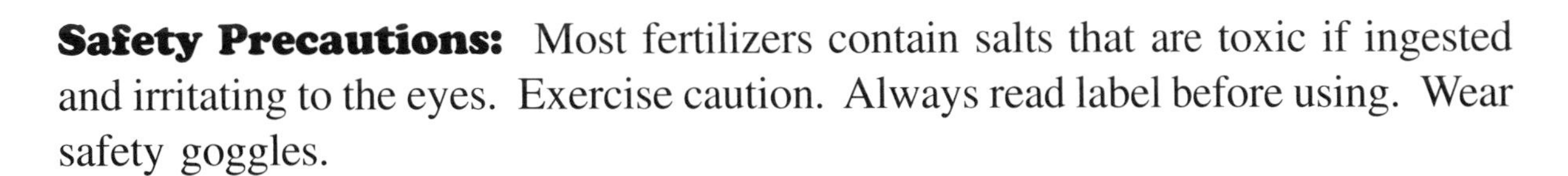

Safety Precautions: Most fertilizers contain salts that are toxic if ingested and irritating to the eyes. Exercise caution. Always read label before using. Wear safety goggles.

Procedure:

1. Obtain four plastic cups, and in each place several teaspoons of soil.
2. To each cup, place several teaspoons of distilled water (approximately the same volume as that of the soil).
3. Mix each thoroughly.
4. Record the pH of each soil type.
5. Label each cup according to the type of fertilizer you will add. The first cup should be labeled "control" and should not receive any fertilizer.
6. Add about a half-teaspoon of aluminum sulfate to the second cup and stir thoroughly.
7. Add about a half-teaspoon of lime to the third cup and stir.
8. Add about a half-teaspoon of Miracle-Gro (or equivalent) to the fourth cup and stir.
9. Repeat with any additional fertilizers in separate cups.
10. Record the pH of each cup after the fertilizers have been added.

Explanation: The aluminum sulfate should significantly lower the pH, thus making the soil more acidic. The lime will significantly raise the pH, making the soil more alkaline or basic. Miracle-Gro, as well as many other fertilizers, have little or no effect on soil pH, since they contain nutrients that are essentially neutral.

Some plants require a narrow pH range in order to thrive. Many evergreens such as rhododendrons, azaleas, and pine trees require acidic soil. Many lawn grasses do not do well in a low pH soil, thus the addition of lime to lawns is a common practice in areas with acidic soils.

Acids and Bases – Experiment # 14:

TESTING FOR CALCIUM CARBONATE

Objective: To determine if a substance contains calcium carbonate.

Materials:

- Vinegar
- Plastic cups
- Red cabbage (see Acids and Bases – Experiment # 1)
- Eggshells
- Tums
- Chalk
- Seashells
- Various rock specimens

Safety precautions: None

Procedure:

1. Place each substance to be tested in a small plastic cup.
2. Pour enough vinegar over each to cover.
3. Carefully observe each every 5 minutes for one hour. Leave overnight to make a final observation the following day.
4. The presence of bubbles indicates calcium carbonate is present. In one of the cups where bubbling is observed, add some red cabbage juice. The vinegar will cause it to turn red. By morning, it should be purple.

Explanation: Calcium carbonate, often called lime, is a substance that will neutralize an acid. If a rock contains calcium carbonate, it will bubble if an acid is added, due to the release of carbon dioxide gas. The reaction is as follows:

$$CaCO_{3(s)} + HC_2H_3O_{2(aq)} \Rightarrow CaCH_3COO_{(aq)} + H_2O_{(l)} + CO_{2(g)}$$

Calcium carbonate   Acetic acid   Calcium acetate

Calcium carbonate reacts with vinegar to yield calcium acetate (a salt), water, and carbon dioxide gas. The CO_2 gas is responsible for the bubbling or fizzing that we observe. Chalk, Tums, eggshells, and seashells are primarily composed of calcium carbonate. Limestone is the most common rock that contains calcium carbonate. If any of your rocks fizzed, it was most likely due to the presence of limestone. The acid test is a common test that geologists use to determine the identity of a rock sample. When calcium carbonate reacts with an acid, it neutralizes that acid. Thus, the red cabbage indicator changes from red to purple.

Acids and Bases – Experiment # 15:
HOW WOULD YOU LIKE YOUR TOMBSTONE?

Objective: To determine what substance makes the most weather-resistant tombstone.

Materials:

- Piece of marble or limestone
- Piece of granite
- Plastic cups
- Vinegar

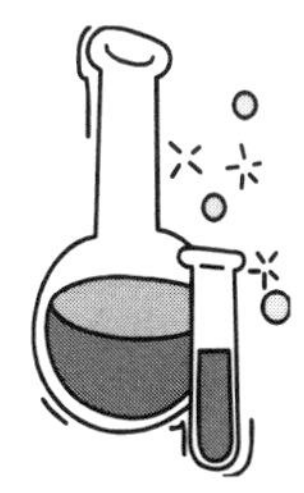

Safety Precautions: None

Procedure:

1. Obtain a piece of marble (or limestone) and a piece of granite.
2. Put each in separate cups and pour over each enough vinegar to cover. The marble will produce bubbles, and the granite will not.
3. Observe until the bubbling stops. Examine the appearance of each.

Explanation: Limestone and marble both contain calcium carbonate, which produces bubbles of carbon dioxide gas in the presence of an acid. Granite, on the other hand, contains no calcium carbonate, and thus is not affected by the acid.

If you visit an old cemetery, you will notice that many of the tombstones are so worn that they are very difficult to read. These are most likely composed of either marble or limestone, and have become severely corroded due to exposure to acidic rainfall for so many years. The same thing is happening to many of the ancient Greek and Roman monuments and statues.

A tombstone made of granite, on the other hand, is resistant to the effects of acidic rainfall, and will remain in excellent condition for many, many years. Granite is such a superior substance that it is used today almost exclusively, as opposed to marble, which was widely used many years ago.

Acids and Bases – Experiment # 16:

BAKING SODA AND BAKING POWDER: WHAT'S THE DIFFERENCE?

Objective: To discover the difference between baking soda and baking powder.

Materials:

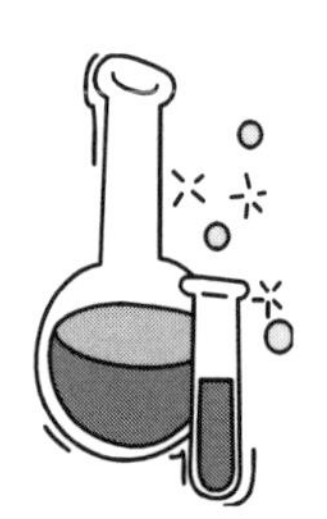

- Baking soda
- Baking powder
- Plastic cups
- pH paper (optional)
- Vinegar
- Eyedropper

Safety Precautions: Wear safety goggles.

Procedure:

1. Place a teaspoon each of baking soda and baking powder in separate cups.
2. Add about a tablespoon of water to each. Observe. Which substance fizzes?
3. If available, test each solution with pH paper.
4. Add a few drops of vinegar to the substance that did not fizz. Observe.

Explanation: Baking soda (sodium bicarbonate) by itself does not react with water. However, it will react vigorously with vinegar or any other acid, giving off carbon dioxide gas, which produces a great deal of fizzing. Baking powder, on the other hand, does react with water, because it is a combination of sodium bicarbonate as well as other ingredients which are acidic. Read the ingredients on the label to verify this. When water is added to the baking powder, the acid then goes into solution and readily reacts with the sodium bicarbonate, giving off CO_2.

If the resulting solutions are tested with pH paper, the baking soda will register about an 8, which is clearly basic. The baking powder, on the other hand, should be closer to neutral, since it contains both basic and acidic ingredients, which neutralize each other.

Baking powder is often used in baking, since it gives off bubbles of CO_2 when mixed with water, thus causing baked goods to rise. If a recipe calls for baking soda, it will also call for an acidic substance to react with, in order to give off CO_2. However, if a recipe calls for baking powder, it will generally not call for another acidic ingredient, since the acid is contained within the baking powder itself.

Acids and Bases – Experiment # 17:

COAGULATION OF MILK

Objective: To create sour milk instantly, by the addition of an acid.

Materials:

- Several transparent plastic cups
- Milk
- Vinegar

Safety Precautions: None

Procedure:

1. Fill the cup halfway with milk.
2. Add to the milk a half-cup of vinegar.
3. Slowly pour into another cup to observe the large clumps that have formed.

Explanation: The vinegar, which is an acid, causes the proteins of the milk to coagulate. This is why sour milk will curdle. When milk sours, the lactose (the sugar in milk) is converted into lactic acid by bacterial action. This lactic acid produces the curds in milk. The liquid part of the milk is known as whey. A process similar to the one used in this experiment is used to make cottage cheese.

Acids and Bases – Experiment # 18:

CARBONIC ACID PRODUCTION

Objective: To produce carbon dioxide gas and observe its reaction with water.

Materials:

- Red cabbage juice (see Acids and Bases – Experiment #1)
- Transparent cup
- Aquarium tubing
- Milk jug
- Drill
- Baking soda
- Vinegar

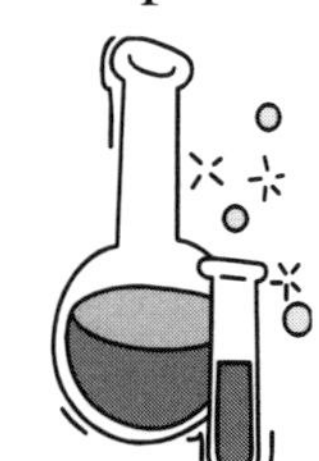

Safety Precautions: Exercise caution when using electric drill. Wear safety goggles while performing this experiment.

Procedure:

1. Measure the diameter of the plastic tubing, and drill a hole in the lid of the milk jug so that the tubing fits in snugly.
2. Insert the tubing through the hole in the lid, and then place the other end of the tubing on the bottom of a cup of red cabbage juice.
3. Place several teaspoons of baking soda in the bottom of the milk jug.
4. Add a half-cup of vinegar and then quickly put on the cap. The CO_2 gas will flow through the tubing into the cup of cabbage juice, changing it from purple to red.
5. Repeat this experiment with other types of acid-base indicators, if available. Bromothymol blue works very well.

Explanation: The CO_2 gas reacts with water to form carbonic acid, which causes the cabbage juice to turn from purple to red. The balanced chemical equation is as follows:

$$CO_{2(g)} + H_2O_{(l)} \Rightarrow H_2CO_{3(aq)}$$

Acids and Bases – Experiment # 19:

DECOMPOSITION OF CALCIUM CARBONATE

Objective: To break down calcium carbonate into calcium oxide, which forms calcium hydroxide when added to water.

Materials:

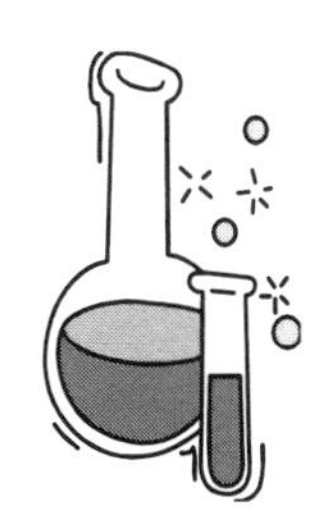

- 2 pieces of chalk (about 1 cm long each)
- Propane torch
- Red cabbage juice (See Acids and Bases – Experiment # 1. Phenolphthalein also works well, if available)
- Two small cups

Safety Precautions: Wear safety goggles. Exercise caution when using propane torch.

Procedure:

1. Place about 30 mL of red cabbage juice in each of two small cups.
2. Place a piece of chalk in one cup of red cabbage juice. There should be no reaction.
3. Heat the other piece of chalk strongly over the flame for five to ten minutes.
4. Place this piece of chalk in the other cup of red cabbage juice. Observe. If there is no reaction, the chalk needs to be reheated.

Explanation: Marble, seashells, chalk, and many other substances contain calcium carbonate ($CaCO_3$), a neutral substance that has no effect on the color of red cabbage juice. When heated, the calcium carbonate breaks down into calcium oxide (CaO) and CO_2. When CaO reacts with water, it forms $Ca(OH)_2$, which is a base. For this reason, the red cabbage juice turns green. The balanced chemical equations are as follows:

$$CaCO_{3(s)} \Rightarrow CaO_{(s)} + CO_{2(g)}$$

$$CaO_{(s)} + H_2O_{(l)} \Rightarrow Ca(OH)_{2(aq)}$$

Acids and Bases – Experiment # 20:

HOW DO COLOR-CHANGE MARKERS WORK?

Objective: To discover the chemistry behind color-change markers.

Materials:

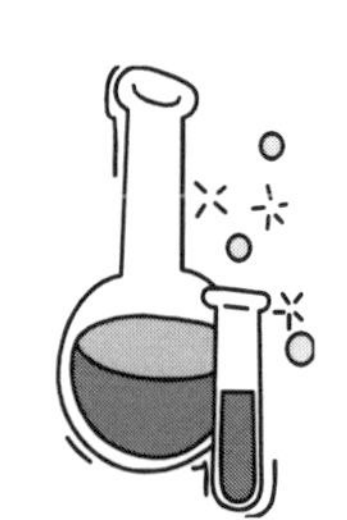

- Crayola Changeables color-change markers
- White paper
- Household ammonia
- Eyedropper
- pH paper (optional)

Safety Precautions: Ammonia is very toxic if inhaled or ingested. Be careful not to inhale fumes. Wear safety goggles.

Procedure:

1. On a white sheet of paper, make a bold line with each of the six markers.
2. Draw over each line with the color-change marker to produce six different colors.
3. Now place a drop of ammonia on each of the original colored lines, a safe distance away from where you drew over each with the color- change marker.
4. You will notice that adding the ammonia produces the same color change as drawing with the color-change marker.
5. If you have some pH paper, draw on a piece with the color-change marker. It will register a pH of around 11 or 12. Now add a drop of ammonia to the pH paper – it will register a similar pH.

Explanation: The color-change marker has a high pH – one that is very basic. Each colored marker in this set contains an alkaline indicator that changes color in the presence of a basic pH. Hence, ammonia has the same effect as the color-change marker, since each has a similar pH.

Acids and Bases – Experiment # 21:

HOW DOES MILK OF MAGNESIA WORK?

Objective: To discover how Milk of Magnesia relieves acid indigestion.

Materials:

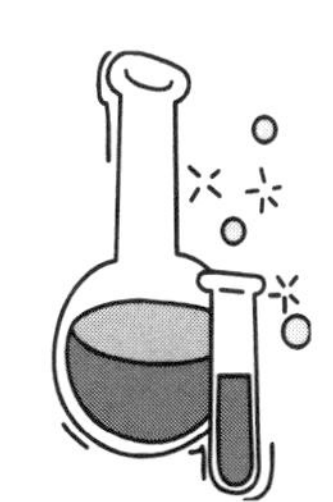

- Milk of Magnesia liquid antacid
- Vinegar
- Tall transparent glass or beaker
- Red cabbage juice (see Acids and Bases – Experiment # 1)

Safety Precautions: Store all medications out of the reach of children.

Procedure:

1. Add a teaspoon of Milk of Magnesia to 100 mL of water in a transparent glass.
2. Slowly add vinegar, stirring as you pour. Add vinegar until the liquid becomes clear.
3. Repeat steps 1 and 2, except this time add some red cabbage juice to the glass before you begin. Observe the color changes.

Explanation: Milk of Magnesia does not dissolve in water. It is a suspension, which must be shaken thoroughly before using. The formula for Milk of Magnesia is $Mg(OH)_2$. Since it is a base, it effectively neutralizes excess stomach acid. Many bases, such as NaOH, cannot be used as antacids, because even though they would neutralize stomach acid, they would severely burn your throat when you swallowed. But Milk of Magnesia can be swallowed without doing any harm to your throat, because it is insoluble in water. However, it is very soluble in acid, which is why it turns clear when an acid is added. It can thus effectively neutralize stomach acid and still be safely swallowed.

When vinegar is added, the two products formed are magnesium acetate ($Mg(C_2H_3O_2)_2$) and water. Magnesium acetate is soluble in water. This reaction is represented by the following balanced chemical equation:

$$2HC_2H_3O_{2(aq)} + Mg(OH)_{2(s)} \Rightarrow Mg(C_2H_3O_2)_{2(aq)} + 2H_2O_{(l)}$$

When the red cabbage juice is added, you can see the color change from green to purple to red as the Milk of Magnesia is neutralized and then excess acid is added.

CHAPTER 8
SOLUTIONS AND SOLUBILITY

A solution is also known as a homogeneous mixture, meaning it has a uniform composition throughout. It is composed of a solute dissolved within a solvent. In a true solution, the solute molecules will never settle out. Salt water, Kool-Aid, and soda pop are all examples of true solutions.

Solubility refers to the ability of a solute to dissolve in a solvent. When something dissolves, the particles become so small that they are rendered invisible. It is for this reason that solutions are always transparent, allowing light to pass through. A substance that does not dissolve in another is termed insoluble.

As the experiments on the following pages will demonstrate, solutions are all around us and solubility affects many aspects of our lives . . .

Solutions and Solubility – Experiment # 1:

MAKING A HOMOGENEOUS MIXTURE

Objective: To make a homogeneous mixture of sugar and water.

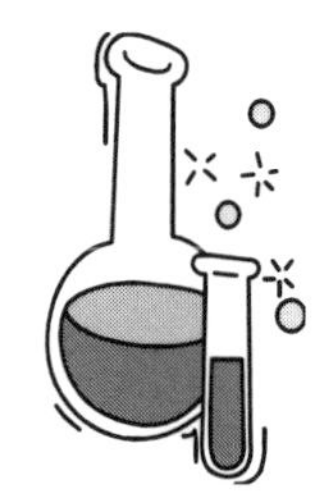

Materials:

- Glass jar with lid
- Sugar

Safety Precautions: None

Procedure:

1. Add several teaspoons of sugar to a jar nearly full of water.
2. Shake thoroughly until all of the sugar has dissolved. Observe.
3. Allow the jar, with the lid on, to remain undisturbed for one week. Observe.

Explanation: The sugar breaks down into particles the size of molecules. This process is known as dissolving. The dissolved molecules of sugar are too small to be seen with the naked eye, thus the sugar becomes invisible. This type of mixture is known as a homogeneous mixture, also known as a solution. Homogeneous means that the particles are uniformly distributed throughout the entire solution. There are no layers, and the sugar will never settle out. After one week, you should have observed that the sugar remains in solution. The molecules of sugar are so small that they will remain in solution indefinitely. The solution will also be transparent, as are all true solutions.

What do you think would happen to the sugar if the jar was left uncovered and the water was allowed to evaporate?

Solutions and Solubility – Experiment # 2:

MAKING A HETEROGENEOUS MIXTURE

Objective: To make a heterogeneous mixture of sand and water.

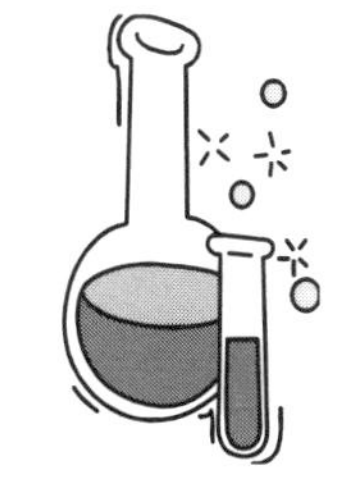

Materials:

- Glass jar with lid
- Sand

Safety Precautions: None

Procedure:

1. Add several teaspoons of sand to a glass jar nearly full of water.
2. Shake vigorously. Observe.

Explanation: Sand and water do not mix. They form what is known as a heterogeneous mixture, also known as a temporary suspension. Heterogeneous means that the mixture is not uniformly distributed, but rather is separated into distinct layers. The sand particles will remain suspended temporarily while the mixture is being shaken, but due to the force of gravity will immediately settle as soon as you stop shaking the jar.

Solutions and Solubility – Experiment # 3:

MAKING A COLLOID

Objective: To make a colloidal suspension of clay and water.

Materials:

- Clay soil
- Glass jar with lid

Safety Precautions: None

Procedure:

1. Obtain some clay soil from your yard or a construction site. This will often be the layer below the topsoil. When moist, it can easily be rolled into a ribbon.
2. Place several teaspoons of the clay soil into a jar nearly filled with water and shake vigorously.
3. Allow to remain undisturbed for one week. Observe.

Explanation: After one week, many of the clay particles will have settled to the bottom. However, some of the clay particles will remain suspended in the water, giving it a cloudy appearance. These particles may remain suspended indefinitely. But the particles have not truly dissolved. If they did dissolve, the mixture would be clear. A mixture composed of particles that are larger than molecules, yet do not settle out, is known as a colloid. A colloid is also known as a permanent suspension. The particles of a colloid remain suspended due to the bombardment of neighboring water molecules, which are constantly in motion. The particles of a colloid typically range in size from 1 to 1000 nanometers (nm). A nm is a billionth of a meter, which is much larger than a molecule.

A colloid can easily be identified by the fact that it has a cloudy appearance, yet does not settle out into layers under normal circumstances. Colloids can be composed of liquids, solids, or gases. Some other common examples of colloids are milk, mayonnaise, paint, butter, cheese, gelatin, and blood.

Solutions and Solubility – Experiment # 4:

WHY WATER AND OIL DON'T MIX

Objective: To discover why oil and water do not mix.

Materials:

- 2-Liter bottle
- Vegetable oil
- Food coloring
- Table salt

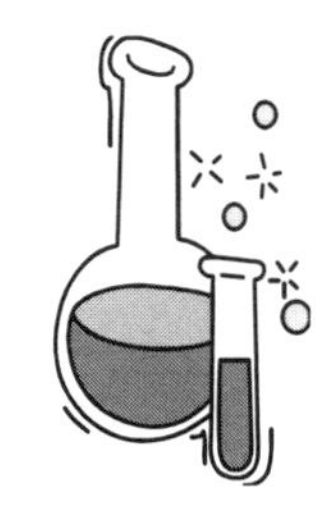

Safety Precautions: None

Procedure:

1. Add about 500 mL of vegetable oil to the 20 oz bottle. Add a few drops of food coloring and a little salt. Observe.
2. Add about 500 mL (about two cups) of water to the bottle.
3. Shake thoroughly. Observe.

Explanation: When the water is added, it will sink to the bottom of the bottle, causing the oil to rise. The oil floats on top because it is less dense than water. The entire contents of the bottle are a good example of a heterogeneous mixture, which contains distinct layers. The oil and water do not mix because water is polar and oil is not. A polar substance is composed of molecules that contain a positive and a negative end. A nonpolar substance does not contain these regions of positive and negative charges. Since opposite charges attract, the positive end of one polar molecule is attracted to the negative end of a neighboring polar molecule. Since these polar molecules are attracted to one another, the oil is pushed out of the way, thus the oil and water do not mix. The principle at work here is "like dissolves like." This means that a polar substance will generally not dissolve in a nonpolar substance. But the oil, a nonpolar substance, will tend to dissolve in other nonpolar substances.

The oil, being nonpolar, will not dissolve any liquids or solids that are polar. Salt and food coloring dissolve in water; therefore these substances must be polar. A nonpolar substance generally cannot dissolve a polar substance. In this experiment, the behavior of the food coloring is particularly interesting – it remains in the shape of perfect spheres.

Solutions and Solubility – Experiment # 5:

DOES SALT REALLY DISSOLVE IN WATER?

Objective: To discover whether or not salt dissolves in water.

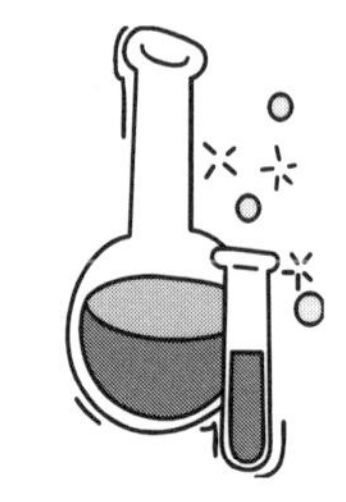

Materials:

- Two glass jars
- Rock salt
- Table salt

Safety Precautions: None

Procedure:

1. Add several teaspoons of table salt to a jar nearly full of water. Shake until all salt has dissolved. Observe the appearance.
2. Add several teaspoons of rock salt to a jar nearly full of water. Shake until all salt has dissolved. Observe the appearance.

Explanation: The mixture of rock salt should be completely clear. Rock salt, which is pure sodium chloride, dissolves very well in water. Sodium chloride is composed of positive sodium ions and negative chloride ions, which are pulled apart by the polar water molecules.

The table salt, on the other hand, forms a very cloudy mixture. This is not due to the sodium chloride, but rather due to other things that are added to the salt. If you read the label on a container of table salt, one of the ingredients may be a silicoaluminate compound. This is a compound that does not dissolve in water, but rather forms a colloid. It is added to table salt to prevent it from clumping up, so the salt will flow freely from salt-shakers. Rock salt, which exists in large clumps, does not contain this added substance.

The Morton salt company was one of the first to use additives to keep the salt in salt shakers freely moving. Before these additives were used, the salt would often clump up on days when there was a lot of moisture in the air. Thus was born the Morton salt slogan, "When it rains, it (the salt) pours."

If you examine ocean water or a salt water aquarium, it will be completely clear, not cloudy!

Solutions and Solubility – Experiment # 6:

LIKE DISSOLVES LIKE

Objective: To discover whether two nonpolar substances will form a homogeneous solution.

Materials:

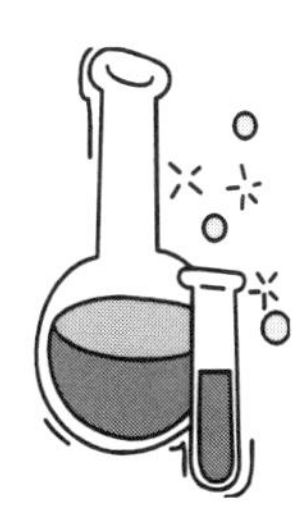

- Kerosene
- Paint thinner
- Glass jars
- Food coloring

Safety Precautions: Kerosene and point thinner are both extremely flammable – keep away from flames. Do not inhale fumes – use only in a well-ventilated area or outside. Both substances are toxic if ingested.

Procedure:

1. Fill a glass jar about one-quarter full with water. Add a few drops of food coloring. Add a little kerosene to the jar. Observe.
2. Fill another glass jar one-quarter full with water. Add a few drops of food coloring. Add a little paint thinner. Observe.
3. Combine the contents of the two jars. Observe.

Explanation: Neither kerosene nor paint thinner will dissolve in water. They form heterogeneous mixtures, separating into distinct layers. However, kerosene and paint thinner together form a homogeneous mixture. Since they are both nonpolar, they dissolve in each other. Since "like dissolves like," a nonpolar substance will tend to dissolve in other nonpolar substances. There are many practical applications of this principle.

DDT (dichlorodiphenyltrichloroethane) was widely used in insecticides for many years, but was banned in the U.S. in 1973 because it accumulated in the fatty tissues of both wildlife and humans. DDT is a nonpolar substance, therefore it readily dissolves in fat, another nonpolar substance.

Vitamins A, D, E, and K are fat soluble. As a result, they can build up to toxic levels in the body if excessive amounts are ingested. But a water soluble vitamin, such as Vitamin C, will be harmlessly excreted by the urine if more is ingested than the body needs.

Solutions and Solubility – Experiment # 7:

THE CHEMISTRY OF STAIN REMOVAL

Objective: To discover which substances are most effective at removing stains.

Materials:

- Strips of white cotton cloth
- Used motor oil
- Turpentine
- Paint thinner
- Kerosene
- Lighter fluid
- Alcohol
- Hydrogen peroxide
- Other household substances
- Eyedroppers
- Clear plastic cups

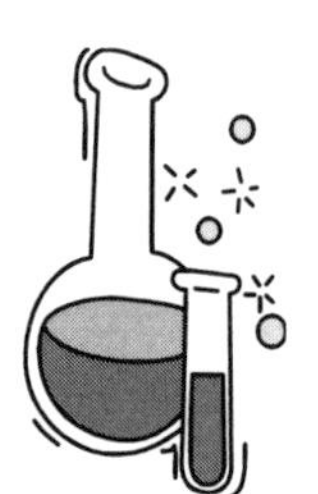

Safety Precautions: Most of the above substances are flammable, emit harmful vapors, and are toxic if ingested. Keep away from flames. Do not inhale vapors. Do this experiment outside or in a well-ventilated area. Wear safety goggles.

Procedure:

1. Place several drops of each of the above substances in individual cups of water to determine if they are polar or nonpolar. If the substance dissolves in water, it is polar. If it does not dissolve in water, it is nonpolar.
2. Place a drop of the used motor oil on separate pieces of the white cloth.
3. Place a liberal amount of each of the above substances on each oil stain, and scrub by folding the cloth over and rubbing together both halves. Rinse with water and allow each to dry.
4. To serve as a control, rinse one stain with water only.
5. Observe the extent of stain removal on each cloth. What conclusions can you make?

Explanation: The motor oil is clearly nonpolar, since it does not mix with water. Although results may vary widely for this experiment, as a rule of thumb, the nonpolar substances are generally the most effective at removing stains. This is because the nonpolar oil is dissolved by the nonpolar solvents, thus removing the stain. Some brands of lighter fluid are even labeled as a tar and grease remover.

Carbon tetrachloride (CCl_4) was formerly widely used as a dry cleaning solvent, due to its nonpolar nature, and was highly effective at removing a variety of stains. However, its use in dry cleaning was discontinued in the mid-1960's due to health and environmental concerns. Carbon tetrachloride is a probable carcinogen, and being nonpolar, is readily absorbed by fat cells in the body.

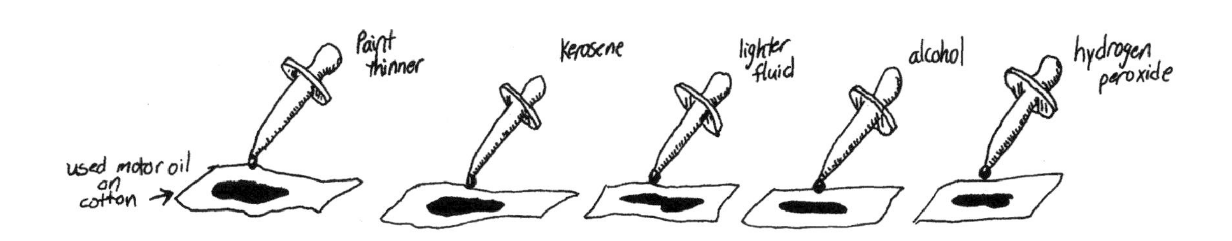

Solutions and Solubility – Experiment # 8:

DIFFUSION OF WATER UP A CELERY STALK

Objective: To demonstrate the process of diffusion using a celery stalk.

Materials:

- Celery
- Transparent drinking glass
- Food coloring

Safety Precautions: None

Procedure:

1. Fill a glass with water and add a few drops of food coloring.
2. Place a celery stalk in the glass.
3. Observe daily for one week.

Explanation: Through diffusion, water moves from a region of high concentration to one of low concentration. The concentration of water is much higher in the glass of water than it is in the celery stalk. Therefore, the water will move up the celery stalk. This is easy to observe, as the top of the celery stalk will be the color of the water in the beaker.

A special type of diffusion, known as osmosis, occurs in this experiment. Osmosis is the diffusion of water through a semipermeable membrane. In cells this selectively permeable membrane that surrounds the cell is known as the plasma membrane, which allows water to pass into or out of the cell. Plant cells also contain a cell wall outside of the cell membrane, which tends to keep plants rigid.

Solutions and Solubilty – Experiment # 9:
MAKING A PICKLE

Objective: To make a pickle by placing a cucumber in salt water.

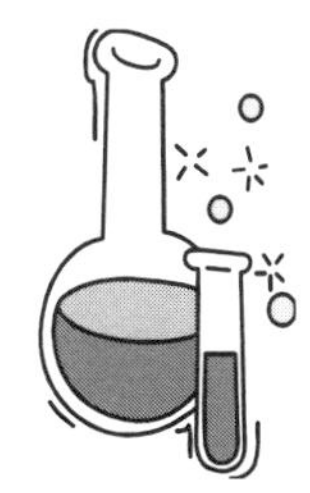

Materials:
- Glass jar with lid
- Whole cucumber
- Salt

Safety Precautions: Do not eat the product of this experiment. The cucumber is not edible, as it not produced under sanitary conditions.

Procedure:
1. Dissolve as much salt as possible in a nearly full jar of water.
2. Place a whole cucumber in the jar and replace the lid.
3. Observe daily for several weeks.

Explanation: Since the water concentration is greater in the cucumber than it is in the salt water, water will tend to move out of the cucumber and into the salt water in an attempt to achieve equilibrium – a state where the concentration of water inside the cucumber is the same as that outside the cucumber. As a result, the cucumber will shrink. A similar process is used to make pickles commercially. As we saw in the previous experiment, the movement of water across a semipermeable membrane is known as osmosis.

Solutions and Solubility – Experiment # 10:

ENLARGING A CUCUMBER

Objective: To enlarge a cucumber by placing it in distilled water.

Materials:

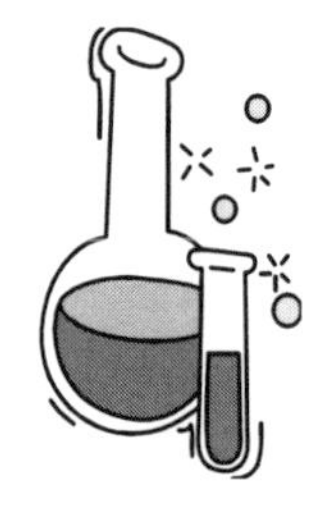

- Glass jar with lid
- Distilled water
- Whole cucumber

Safety Precautions: None

Procedure:

1. Place a cucumber in a jar nearly full of distilled water.
2. Observe daily for several weeks.

Explanation: Water will tend to move from a region of higher concentration to one of lower concentration in an attempt to achieve equilibrium. The concentration of water in the cucumber is less than that of the distilled water surrounding it, due to the presence of dissolved solutes. Therefore, through the process of osmosis, water will flow into the cucumber, causing it to enlarge. In the last experiment, where a pickle was made, water flowed in the opposite direction. In both cases, osmosis is occurring. Osmosis can occur in either direction, like water passing through a screen.

Solutions and Solubility – Experiment # 11:

DOES TEMPERATURE AFFECT THE RATE OF DISSOLVING?

Objective: To determine the effect of temperature on the rate of dissolving.

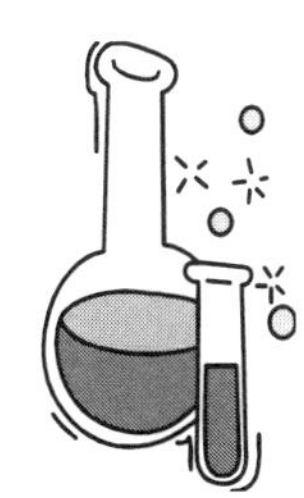

Materials:

- Two transparent cups or beakers
- Sugar cubes
- Microwave or stove

Safety Precautions: Exercise caution when using hot water.

Procedure:

1. Fill a cup with cold water, and then another with hot water.
2. At the same time, drop a sugar cube in each.
3. Record the time it takes for each to dissolve.

Explanation: Substances clearly dissolve much faster in hot water than in cold water. The molecules of hot water are moving much faster. Therefore, there are many more collisions of hot water molecules with the sugar cube, causing it to dissolve much more quickly. The more collisions between solvent (water) and solute (sugar) molecules, the faster the rate of dissolving.

Solutions and Solubility – Experiment # 12:

DOES SURFACE AREA AFFECT THE RATE OF DISSOLVING?

Objective: To determine if surface area affects the rate of dissolving.

Materials:

- Two transparent drinking glasses
- Sugar cubes

Safety Precautions: None

Procedure:

1. Crush a sugar cube into fine granules.
2. Simultaneously drop the crushed sugar cube and an intact sugar cube into separate glasses of water.
3. Record the time it takes for each to completely dissolve.

Explanation: The surface area of a substance plays a major role in its rate of dissolving. It is the collision of water molecules against the sugar particles that cause them to break up and dissolve. If more of the sugar is in contact with the water, then more collisions can take place. With the intact sugar cube, only the outside surfaces are initially in contact with the water, so it dissolves more slowly.

Solutions and Solubility – Experiment # 13:

DOES SHAKING AFFECT THE RATE OF DISSOLVING?

Objective: To determine the effect of shaking on the rate of dissolving.

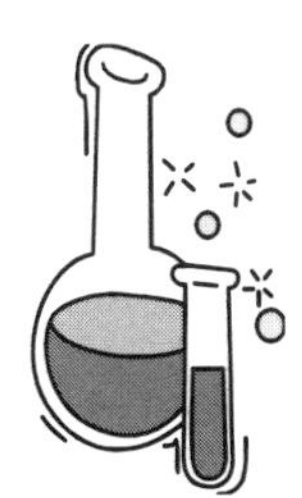

Materials:

- Two glass jars with lids
- Sugar

Safety Precautions: None

Procedure:

1. Place several teaspoons of sugar in each of two jars nearly full of water.
2. Shake one jar vigorously for one minute. Allow the other to remain undisturbed.
3. Compare the amount of undissolved sugar in each jar.

Explanation: The jar of water that was shaken dissolved the sugar much more quickly than the water that was undisturbed. Shaking brings the sugar molecules into contact with more water molecules, increasing the number of collisions between the water and the sugar. As the number of collisions increases, the rate of dissolving increases.

Solutions and Solubility – Experiment # 14:

MAKE YOUR OWN HANDWARMER

Objective: To make a hot pack by dissolving calcium chloride in water.

Materials:

- Calcium chloride (available in hardware stores as an ice melter; often marketed as "Driveway Heat")
- Plastic sandwich bag

Safety Precautions: Calcium chloride will sting if it gets in your eyes. Wear safety goggles. May be toxic if ingested.

Procedure:

1. Place three heaping teaspoons of calcium chloride in a sandwich bag.
2. Add about 50 mL of water.
3. Seal the bag. Knead with your fingers through the bag until most of the calcium chloride salt has dissolved.
4. Note the change in temperature of the water.

Explanation: This experiment is an excellent example of an exothermic process – one where heat is released. When calcium chloride is dissolved in water, energy is required to break the bonds of the calcium and chloride ions that are bonded to one another. However, as the calcium and chloride ions bond to the water molecules, heat is released. This heat that is released when these new bonds form is far greater than the heat that was initially absorbed to break the bonds of the solid calcium chloride. As a result, heat is released when this substance is dissolved in water. Can you think of any other substances that release heat when dissolved in water?

Solutions and Solubility – Experiment # 15:

MAKING MERINGUE: A COLLOID

Objective: To form meringue, which is a type of colloid.

Materials:

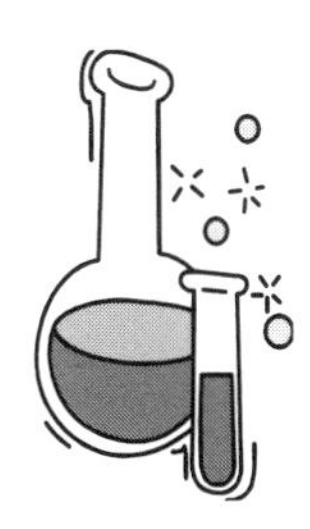

- One egg
- Baking soda
- Measuring cup
- Citric acid (available from a pharmacy or grocery store)
- Large bowl

Safety Precautions: The meringue made in this experiment should not be eaten, since it is made from raw eggs that may be contaminated with Salmonella bacteria. Wash hands thoroughly after doing this experiment.

Procedure:

1. Place an egg white in a large mixing bowl.
2. Add a half-cup of water and two teaspoons of baking soda.
3. Add two teaspoons of citric acid and mix thoroughly. Observe.

Explanation: Because meringue is filled with a gas, which is less dense than liquids or solids, the volume of meringue produced in this experiment is much larger than the original volume of materials used. Meringue is an example of a foam, which is a type of colloid. A foam is formed when a gas is dispersed in a liquid. In this case, carbon dioxide is formed when the citric acid reacts with baking soda. This gas is trapped inside of the egg white, causing it to expand into a foam. The normal way to make meringue is to introduce air into a mixture of egg whites, sugar, and vanilla by beating.

Solutions and Solubility – Experiment # 16:

MAKING ROCK CANDY

Objective: To make rock candy due to the crystallization of sugar.

Materials:

- Stove or hot plate
- Sugar
- Pan
- Heat-resistant glass jar
- String
- Paper clip
- Pencil
- Wooden spoon for stirring

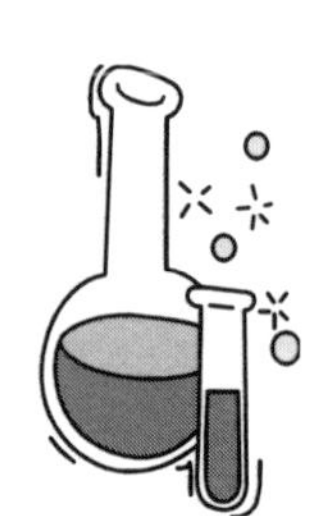

Safety Precautions: Exercise caution when using stove and when pouring hot liquids. The sugar solution will be very hot!

Procedure:

1. To determine the amount of water to use, fill the glass jar and pour into the pan.
2. Gently heat the water, and continually add sugar as long as it is dissolving. The sugar is no longer dissolving when it begins to settle on the bottom. Stir constantly.
3. As the water warms, it will be able to dissolve more sugar. You will be surprised at how much sugar can be added, but continue doing so until the water boils.
4. Once the water boils, add no more sugar. Allow to boil for approximately one minute, then remove from the heat and pour into the jar.
5. Tie the string to the pencil and suspend a paper clip from the end of the string. Place in the jar, making sure that the paper clip does not touch the sides or the bottom of the jar.
6. Allow to remain undisturbed for several weeks. As a crust forms on top, break it up daily. This will allow the water to evaporate. As this happens, the sugar will go out of solution and begin to adhere to the paper clip. These crystals will continue to form until all of the water has completely evaporated. The end result is an edible treat!

Explanation: As the sugar dissolves in the water, it forms a saturated solution. As the water is heated, more sugar crystals dissolve, forming a supersaturated solution. A supersaturated solution contains so much dissolved solute that the saturation point of the solution has been exceeded. However, this a fragile state, and any disturbance will force some solute out of solution. As the water evaporates, the remaining water cannot hold as much sugar, so it precipitates out of solution and crystallizes onto the paper clip and string.

A crystal is a solid that forms in a regular repeating pattern. There are several different types of crystal patterns. Examination under a magnifying glass will aid in identifying the particular type of crystal shape. What geometric shape do sugar crystals exhibit?

Related Experiments: The same procedure outlined above can be tried with a variety of substances. Good results can generally be obtained with table salt, Epsom salts, and alum. (However, crystals made from these solutions will not be edible.) Experiment with different substances to see which will form the best crystals.

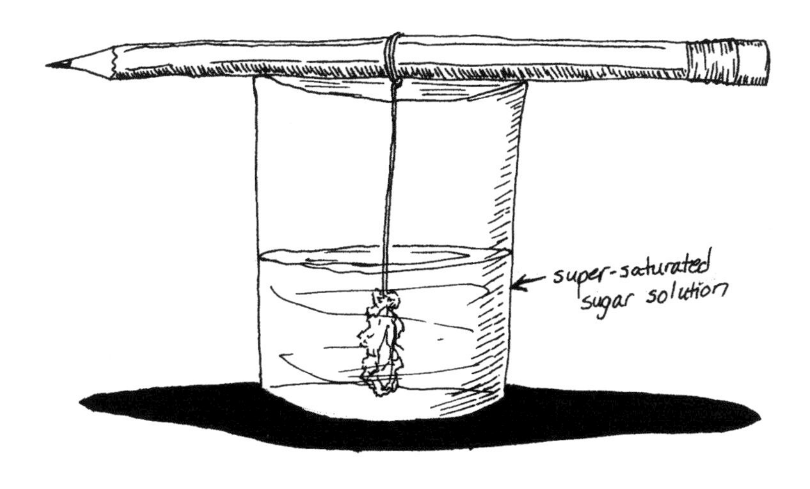

Solutions and Solubility – Experiment # 17:
MAKING ICE CREAM

Objective: To make ice cream in a baggie.

Materials:

- Gallon freezer bag
- Two quart freezer bags
- Rock salt
- Crushed ice
- Measuring spoons
- Milk
- Evaporated milk
- Sugar
- Vanilla extract

Safety Precautions: Wear gloves to prevent your hands from getting too cold.

Procedure:

1. In the quart freezer bag, place the following: ½ cup milk, ¼ cup evaporated milk, 4 teaspoons sugar, and ½ teaspoon vanilla extract. Seal the bag.
2. In the gallon bag, fill halfway with ice. Add 1 cup rock salt.
3. Place the quart bag in the gallon bag and then seal the gallon bag. Knead with your hands for about 20 minutes to a half-hour. After this time, remove the small bag. It should be frozen into ice cream. If not, seal both bags and knead for a few more minutes. Then enjoy a delicious treat!

Explanation: The salt melts the ice by breaking apart the hydrogen bonds that form its hexagonal crystal lattice structure. Melting is an endothermic process, which means energy is absorbed from the surroundings. In this case, the surroundings are comprised of the ice cream mixture in the quart bag. As energy is removed from this mixture, it will freeze into ice cream. Kneading is important to the process, since this introduces air into the mixture, giving ice cream its distinctive taste and texture.

When salt is added to water, it produces a freezing point depression in the water, substantially lowering the temperature at which water will freeze. To make ice cream, it is necessary to have temperatures lower than the normal freezing point of water (0°C), because the addition of sugar, vanilla, and the other substances used in making ice cream depress the freezing point of water. Therefore a temperature below 0°C is required in order for the mixture in the quart bag to freeze into ice cream. The addition of salt to ice is one way to easily produce temperatures below 0°C.

Solutions and Solubility - Experiment # 18:

MAKE YOUR OWN COLD PACK

Objective: To make a cold pack by dissolving ammonium nitrate in water.

Materials:

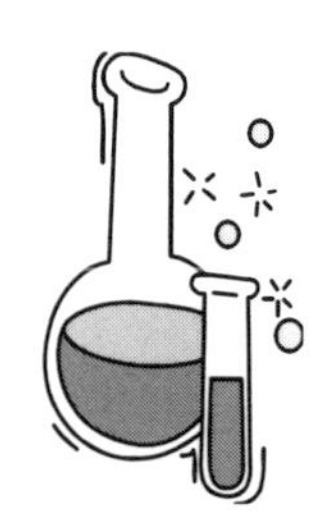

- Ammonium nitrate (available as 34-0-0 fertilizer from an agricultural supply company)
- Plastic sandwich bag

Safety Precautions: Ammonium nitrate is toxic if ingested, and irritating to the skin and eyes. Wear safety goggles.

Procedure:

1. Place 3 heaping teaspoons of ammonium nitrate (NH_4NO_3) in a sandwich bag.
2. Add 50 mL of water.
3. Seal the bag. Knead thoroughly with your fingers through the bag until most of the solid has dissolved. Note the temperature change.

Explanation: The commercially available cold packs also utilize ammonium nitrate. This is a good example of an endothermic process, where energy is absorbed from the surroundings. In order to break apart the bonds of the solid ammonium nitrate particles, energy must be absorbed. Energy is then released when the NH_4^+ and NO_3^- ions bond to the water after dissolving. However, it takes more energy to break apart the ionic solid than is released when the dissolved ions bond to the water. Therefore the overall process is endothermic, since energy is absorbed from the surroundings. If your hands become the surroundings, then your hands will feel cold because the cold pack is removing energy – not adding coldness – to your hands. There is really no such thing as cold. What we commonly refer to as cold is simply the lack of energy.

Solutions and Solubility – Experiment # 19:

SEPARATION OF PIGMENTS USING CHROMATOGRAPHY

Objective: To identify the different types of pigments found in black ink.

Materials:

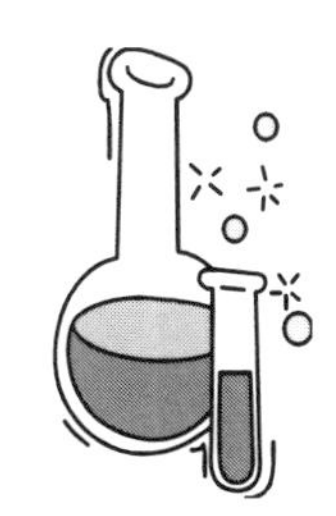

- Paper towels
- Cup
- Variety of black, water-soluble markers

Safety Precautions: None

Procedure:

1. Tear a paper towel into strips approximately 1 inch wide.
2. About an inch from the bottom of the strip, make a large dot with a black marker.
3. Place about a half-inch of water in a cup, and position the strip of paper towel in the water so that the black dot is above the water level. Fold the rest of the paper towel over the top of the cup so it remains in place.
4. Remove the strip of paper towel when the water has traveled to near the top of the strip. Observe.

Explanation: You should observe a clear separation of the different colored pigments that actually make up black ink. The separation of the components of a mixture in this way is known as chromatography. In the experiment you performed, there exists a mobile phase and a stationary phase. The stationary phase is the paper, and the mobile phase is the water. Water tends to move up the paper towel through the process of capillary action. Capillary action refers to the tendency of a liquid to move upward through small tubes or pores, like those found in paper towels. This process is also known as wicking, and greatly increases the absorbency of paper towels.

The pigments that have the greatest attraction to the stationary phase (the paper) will separate out first. The pigments that separate out first will also have the least attraction to the mobile phase (the water). The pigments that separate out last will have the least attraction to the stationary phase, and the greatest attraction to the mobile phase. This difference in attraction is due to differences in solubility, polarity, and molecular mass of the pigments.

Solutions and Solubility – Experiment # 20:

RADIAL CHROMATOGRAPHY

Objective: To produce radial patterns on a circular piece of paper using chromatography.

Materials:

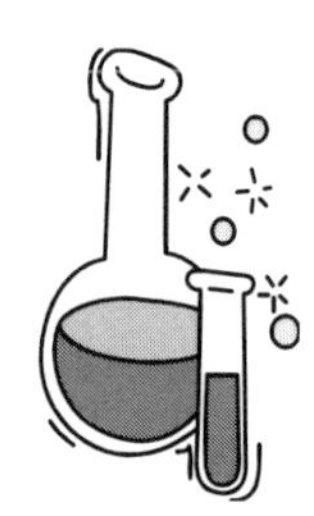

- Coffee filters (or circular filter paper if available)
- Push pin
- Scissors
- Black, water-soluble markers
- Paper cup

Safety Precautions: None

Procedure:

1. Cut away the edges from a coffee filter so you are left with a circular disk.
2. Make a small hole in the center of the disk with a push pin.
3. Make a circle of several black dots around this hole with several different brands of black markers.
4. Cut out a piece of the discarded portion of the coffee filter into a pie-shaped wedge. Twist this into a tight circle, so one end is pointed.
5. Insert the pointed end of the wedge into the hole in the center of the circular disk. It should fit firmly into the hole.
6. Fill a paper cup nearly to the top with water. Make sure the rim of the cup is dry. Place the circular disk on top of the cup so that the large end of the wedge is below the water level. The wedge will act as a wick to draw water through the circular disk.
7. When the water has nearly reached the edge of the disk, remove from the cup. Observe.

Explanation: Beautiful radial patterns can be observed on the coffee filter as the pigments of ink are separated through chromatography. Experiment with different types of markers and different patterns on the filter paper. Try drawing different designs on the paper. Beautiful results can be obtained with a little creativity.

CHAPTER 9
POLYMERS

Polymers are everywhere. Garbage bags, Styrofoam cups, milk jugs, toothpaste tubes, shoe strings, pop bottles, sandwich bags, egg cartons, thread spools, diapers, insulation, caulking, shrink wrap, bubble wrap, Silly String, Silly Putty, Slime, Super Glue, Teflon, toothbrushes, combs, carpet, telephones, floor tile, rubber soles, tires, sunglasses, contact lenses, panty hose, umbrellas, Nerf balls, Frisbees, wet suits, volleyballs, racquetballs, tennis balls, tennis racquets, guitar strings, balloons, rubber bands, credit cards, computers, false teeth, hearing aids, lunch trays, lawn chairs, Astroturf, Velcro, Spandex, footballs, hockey pucks, buttons, erasers, wigs, surfboards, parachutes, sailboats, playing cards, clarinets, flutes, records, videotapes, computer discs, luggage, flea collars, life rafts, pacifiers, baby bottles, photographic film, mannequins, book bags, bowling balls, knapsacks, fishing line, vitamin capsules, sponges, tents, windshield wipers, bubblegum, tents, house paint, toilet seats, Creepy Crawlers, cameras, polyester, bulletproof vests, Band-Aids, rain coats, flags, and puppets are all made from polymers.

What is a polymer? A polymer is a giant molecule composed of many repeating units known as monomers. The prefix poly- means many. Each monomer within a polymer may be identical, or they may be different. A single polymer molecule may be composed of thousands of monomers that are bonded together.

Polymers may be natural, such as starch, proteins, cellulose, and rubber from rubber trees. Plastics, synthetic rubber, and nylon are examples of manmade polymers.

You will begin to explore the wonderful world of polymers in the following chapter . . .

Polymers – Experiment # 1:

WHAT DO THOSE RECYCLING CODES MEAN?

Objective: To understand the meaning of the recycling codes found on many plastic containers.

Materials:

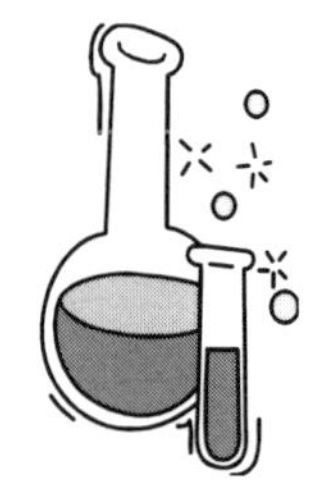

- Various plastic containers found in the kitchen, garage, or bathroom.
- Vegetable oil
- Scissors

Safety Precautions: None

Procedure:

1. Gather as many plastic containers as you can that have a recycling code on the bottom, which is a triangle of arrows with a number in the center.
2. The recycling codes will include numbers from 1-6. Put all containers of the same number in groups, and observe their properties. Note the following:
    a. Translucent, transparent, or opaque?
    b. Uses of this plastic?
    c. Flexible or brittle?
    d. Color?
    e. Flimsy or durable?
    f. Other properties?
3. Cut one container from each category into small pieces and test the density of each by seeing if they float or sink in water. If they sink in water, test them to see if they will float or sink in vegetable oil. Arrange the plastics in order from greatest to lowest density.

Explanation: The purpose of the numbers on the bottom of most plastic containers is to facilitate the recycling of these materials, since most types of plastic can be recycled. The code tells us the type of plastic the container is composed of. The composition of each is as follows:

1 – PETE or polyethylene terephthalate

2 – HDPE or high density polyethylene

3 – PVC or polyvinyl chloride

4 – LDPE or low density polyethylene

5 – PP or polypropylene

6 – PS or polystyrene

For more information about each of the above polymers, visit the Hands On Plastics website at www.handsonplastics.com. In addition to a great deal of background information about polymers, the site offers complete instructions for many great hands-on activities. You can also order a free kit containing a variety of recycled resins and other products that will greatly aid you in your study of polymers.

Polymers – Experiment # 2:

COLLAPSING A STYROFOAM CUP

Objective: To collapse a Styrofoam cup using acetone.

Materials:

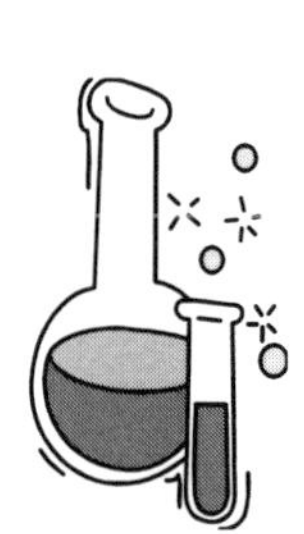

- Acetone (available in hardware stores, or as fingernail polish remover)
- Small aluminum pie pan
- Styrofoam cups
- Wooden popsicle stick

Safety Precautions: Read all warnings on the container of acetone before proceeding. Wear safety goggles. Acetone is very flammable – keep away from open flames. Inhaling acetone vapors is harmful – do this experiment outdoors or in a well-ventilated area. Acetone is toxic if ingested. Do not allow the acetone to contact your skin. Wash hands thoroughly when finished with this experiment.

Procedure:

1. Fill a small aluminum pie pan halfway with acetone.
2. Place a Styrofoam cup open-end down into the acetone, and push down gently with the wooden stick.
3. Remove with the stick and rinse thoroughly with water. You may then fashion it into a ball or other shape.
4. Repeat this procedure with packing peanuts, egg cartons, or other large pieces of Styrofoam.

Explanation: The Styrofoam cup is composed of polystyrene, as evidenced by the recycling code of "6" on the bottom. (See experiment # 1 in this chapter.) Polystyrene is a polymer composed of many styrene monomers bonded together. Acetone will cause the cup to collapse, but not dissolve. The acetone breaks some of the bonds that give the cup its shape, but the end result is still polystyrene. This is a good example of a physical change, since the chemical composition of the substance is not altered.

Styrofoam is filled with air, which enhances its insulating abilities. When the cup is placed in acetone, bubbles can be observed as this air is being released. This removal of air greatly reduces the volume of the cup. When it dries, the polystyrene will again become rigid and maintain its shape.

Many other solvents will also have this effect on Styrofoam. If you have ever tried to spray paint Styrofoam, you were probably disappointed by the corrosive effect of the spray paint. Similar dismay has also been experienced by those who have tried to pour gasoline into a Styrofoam container!

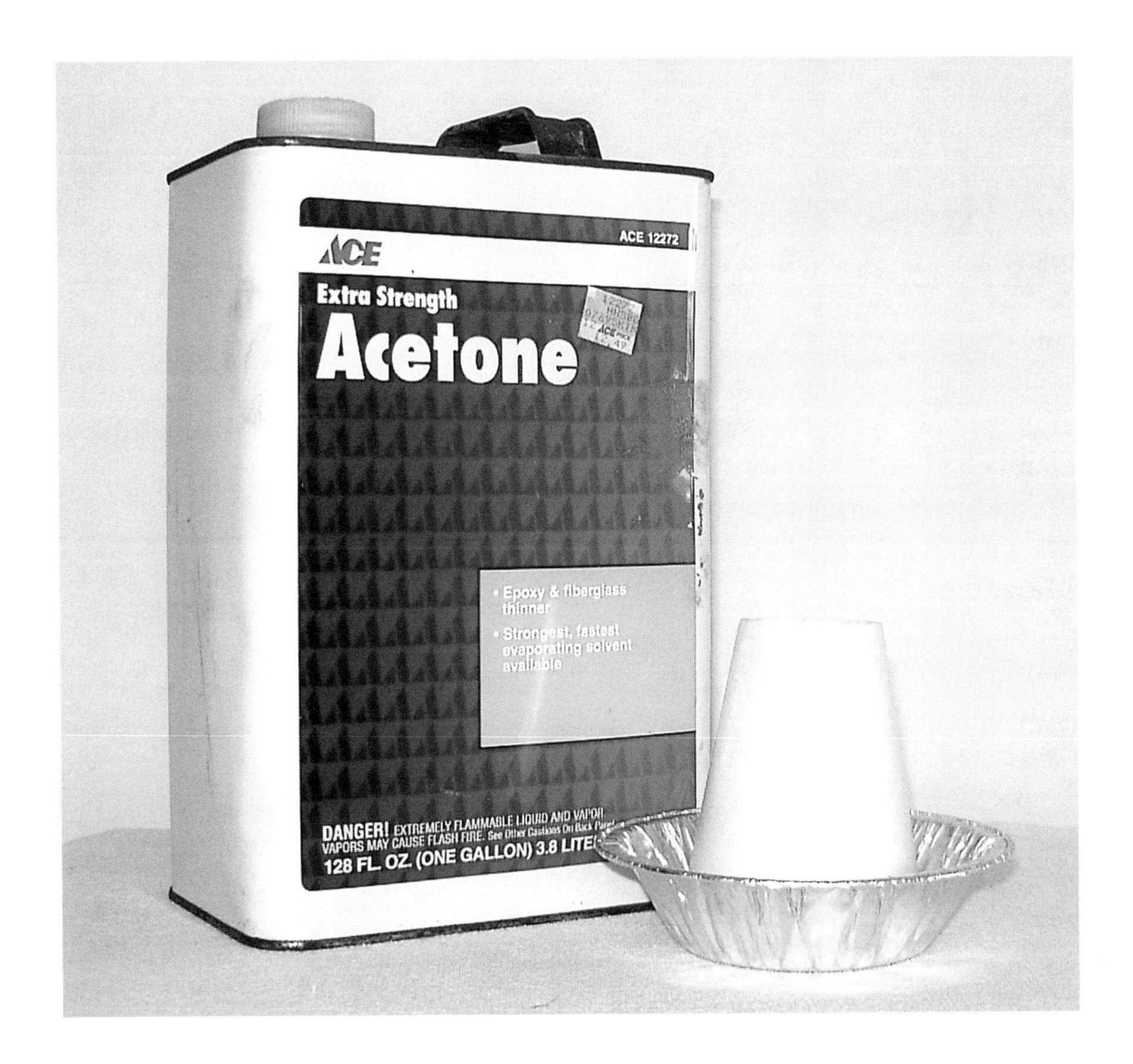

Polymers – Experiment # 3:

FUN WITH PACKING PEANUTS

Objective: To observe the effect of water on starch packing peanuts.

Materials:

- Starch packing peanuts (available from an office supply store)
- Plastic cups
- Tincture of iodine

Safety Precautions: Tincture of iodine is toxic if ingested. Wear safety goggles.

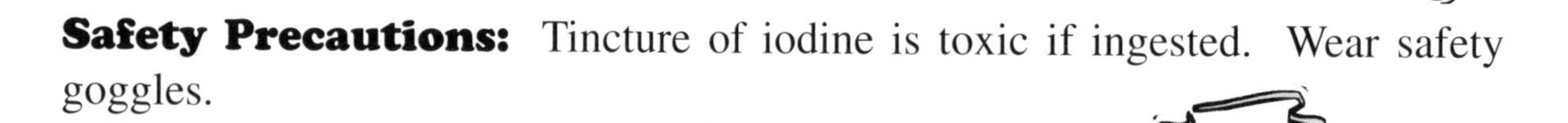

Procedure:

1. Place some starch packing peanuts in a cup half-filled with water. Stir with a spoon. What happens? How many starch packing peanuts can fit into this half-cup of water?
2. Add a few drops of tincture of iodine. Observe.

Explanation: Starch packing peanuts are an excellent alternative to Styrofoam, in that they are completely biodegradable, and thus do not take up valuable space in landfills. When blown up with air, they provide excellent packing material. They do not really dissolve in water, but instead form a thick paste, somewhat like mashed potatoes. The decrease in volume occurs when water collapses the structure of the packing peanuts, allowing the air to be released. The mixture turns black when iodine is added, due to the formation of the starch-iodine complex. (See Chemical Reactions – Experiment # 24.)

Polymers – Experiment # 4:
POKING A NEEDLE THROUGH A BALLOON

Objective: To pass a needle through both sides of a balloon without the balloon breaking.

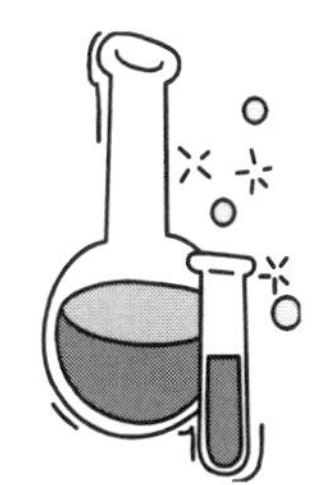

Materials:
- Large furniture needle or wooden bamboo skewer
- Helium-quality balloon
- Vaseline

Safety Precautions: If you are not careful, the balloon may pop!

Procedure:
1. Lubricate the end of the needle with Vaseline.
2. Inflate a balloon to about half of its potential volume. Tie off the balloon.
3. Slowly pass the needle through the thick part of the balloon at its bottom, and then pass it through the balloon and out the other end, near to where it is tied. If done properly, the balloon will not pop!

Explanation: The balloon is made of rubber, which is a polymer. Polymers are composed of long chains of repeating units. The needle slips through and between these polymer chains, much like a needle might slip through a bowl of cooked spaghetti noodles without ever actually piercing a noodle itself. The polymer chains will tend to close back around the entry point of the needle, which will prevent the balloon from deflating for a short time. This little trick does take practice, however, and you may pop a few balloons before you have mastered the technique.

Polymers – Experiment # 5:

MAKE YOUR OWN SLIME

Objective: To make a stretchy, slimy polymer.

Materials:

- Elmer's white glue (or equivalent)
- 20 Mule Team Borax (available from grocery store)
- Food coloring
- 2-Liter bottle
- Eyedropper
- Popsicle stick
- Ziploc storage or freezer bag
- Disposable plastic cups

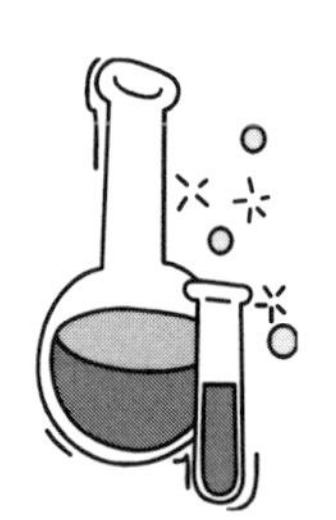

Safety Precautions: Borax is toxic and irritating to the skin. Wear safety goggles.

Procedure:

1. Pour about a quarter-cup of glue into a plastic cup.
2. Add a quarter-cup of water to the glue and stir thoroughly. You now have a 50:50 glue-water mixture. You may make as much or as little slime as you desire, as long as you maintain the 50:50 ratio between the glue and water.
3. Add a few drops of food coloring, if desired, and stir thoroughly.
4. Add a cup of borax powder to an empty 2-L bottle. Fill with water, and shake vigorously for about 10 minutes. This will form a saturated borax solution. All of the borax will not dissolve right away, but most will dissolve over time. You can still use the solution even though all of the solid has not dissolved.
5. Pour a little of the borax solution into an empty cup, and add a little at a time with your eyedropper to the glue-water mixture. The slime will collect on your stick as you stir.
6. It is important that you not add too much borax solution, or the slime will become too stiff. A good rule of thumb is to stop adding borax when there is still a little fluid left in the bottom of the cup. This way, you will not add too much.

7. Remove the slime from the stick with your fingers, rinse off with water, and then work it with your hands.
8. Store in a sealed Ziploc bag.

Explanation: Slime is a fascinating substance that can provide many hours of slimy fun. It is similar in composition to some of the commercial varieties of slime available in stores.

Elmer's glue is primarily composed of the polymer polyvinyl acetate. It has a greater viscosity (resistance to flow) than many liquids, but is not nearly as viscous as slime. The borax solution causes the polyvinyl acetate molecules to become crosslinked. Crosslinking can be compared to the placement of wooden ties to hold the rails of a railroad track in place. The borax (sodium tetraborate) molecules act like the ties, which hold the rails of the polyvinyl acetate molecules in place. This makes for a much more viscous solution, since the polyvinyl acetate and sodium tetraborate molecules are now linked firmly together.

Slime is an example of a non-Newtonian fluid. According to Isaac Newton, the viscosity of a liquid is dependent only on its temperature. But the viscosity of a non-Newtonian fluid, such as slime, can be altered in several other ways besides changing its temperature. If you pull the slime slowly apart, it will form long thin strands. But if pulled apart rapidly, it breaks. It can bounce somewhat if formed into a ball and dropped on a hard surface. If poked quickly with the finger, your finger will bounce off. But if poked slowly with a finger, the slime can be easily deformed. Other examples of non-Newtonian fluids are quicksand and Silly Putty.

Place your slime on a newspaper and then press down firmly. Is the newsprint transferred to the slime? Is your slime a solid or a liquid? If left on the table, what happens to its shape? If your slime assumes the shape of its container, does this make it a solid or a liquid?

Polymers – Experiment # 6:

TESTING THE ABSORBENCY OF DIAPERS

Objective: To test different brands of diapers to discover which is the most absorbent.

Materials:

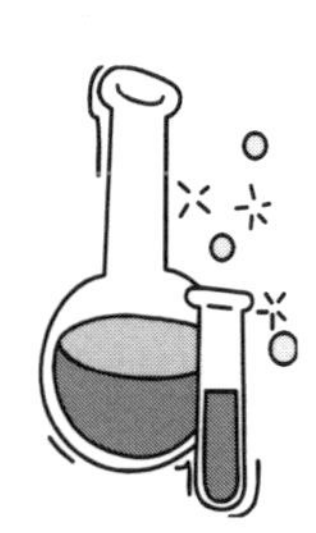

- Several diapers of different brands (all should be roughly the same size)
- Several 2-L bottles
- Distilled water
- Scissors

Safety Precautions: Sodium polyacrylate is toxic if ingested.

Procedure:

1. Using scissors, cut away the top from several bottles so a diaper can easily fit into each. Fill each bottle with about 1.5 Liters of distilled water. Tap water can be used if distilled water is not available.
2. Completely submerge each diaper in a bottle of water, and allow to remain submerged for 5 minutes.
3. Remove each diaper, allowing the excess water to run off back into the container. Do not wring out the diapers.
4. Record the amount of water each diaper absorbed.
5. Rank the diapers in order from most absorbent to least absorbent.
6. Carefully dissect each diaper with scissors by cutting away the lining with scissors. Observe the contents.

Explanation: If you have ever put a baby in a swimming pool with a diaper on, you have already experienced the amazing superabsorbent properties of diapers. All diapers contain a superabsorbent polymer known as sodium polyacrylate, which is capable of absorbing 800 times its weight in distilled water, 300 times its weight in tap water, and 60 times its weight in salt water. Since urine is salty, sodium polyacrylate cannot absorb as much urine as freshwater. If your diaper absorbed 800 mL of distilled water (equal to 800 grams of water) then the diaper contains approximately 1 gram of sodium polyacrylate. Only a very small amount of powder is needed. A very large diaper may contain up to 12 grams of superabsorbent powder, with smaller diapers containing less. When the diaper is dissected, it can be observed that the powder works by absorbing water to form a gel.

Proctor and Gamble introduced the first disposable diaper in 1961, with the advent of Pampers. Sodium polyacrylate was not added to the disposable diaper until 1984. This innovation made diapers thinner, more leakproof, and about 50% lighter.

Sodium polyacrylate has recently been marketed as a substance called Barricade, which is used to prevent the spread of fires. When spread over the surface of a wall or roof, saturated sodium polyacrylate provides an excellent barrier to the spread of fire. It has also been used to stop the spread of forest fires. This application was first suggested by firefighters who noticed that wet diapers would not burn.

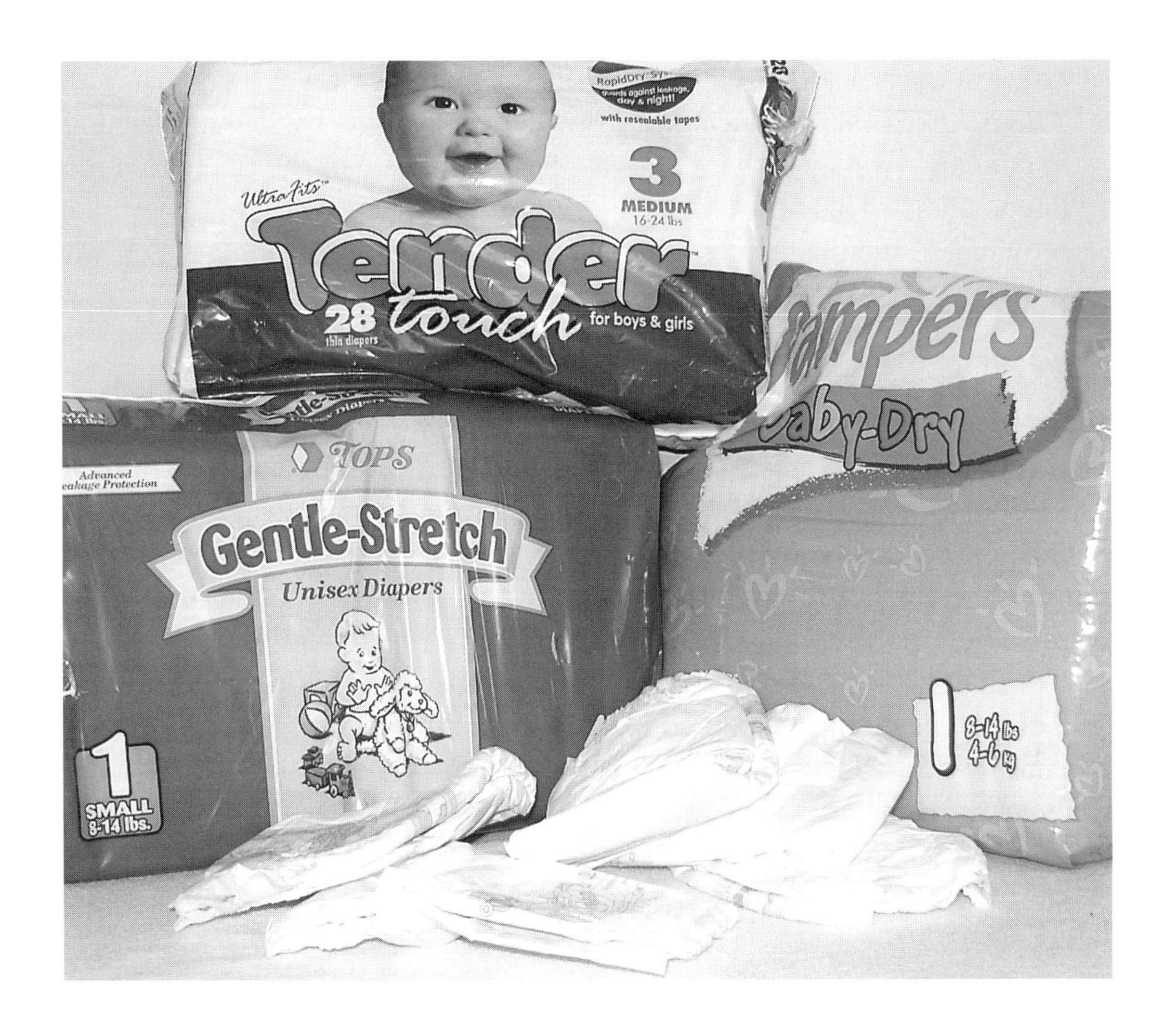

Polymers – Experiment # 7:
A SUPERABSORBENT POLYMER

Objective: To discover the properties of sodium polyacrylate, the active ingredient used in diapers.

Materials:

- Disposable diaper
- Transparent plastic cup
- Distilled water
- Salt
- Scissors
- Popsicle stick

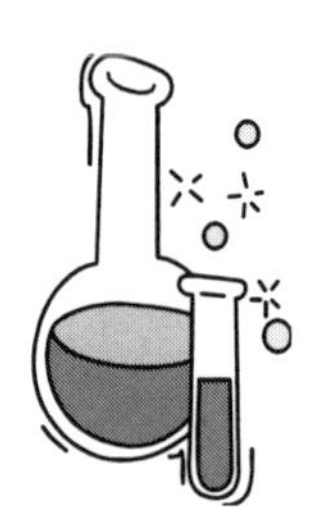

Safety Precautions: Sodium polyacrylate is toxic if ingested.

Procedure:

1. Using scissors, carefully cut away the inner lining of a diaper.
2. Attempt to extract as much of the powder as possible, by shaking the diaper into a cup. (It may be necessary to shake out some of the fibers to remove the powder.) This powder is sodium polyacrylate.
3. Slowly add distilled water and observe as a gel forms. How much water will the powder absorb?
4. Add a couple of teaspoons of salt to the mixture and stir. What happens?

Explanation: The powder you removed from the diaper is what makes the diaper so absorbent (see previous experiment). Salt will draw water from the gel and cause it to turn back into a liquid, because the water within the gel will attempt to equalize the concentration of water outside the gel. In so doing, water is drawn from inside the gel, destroying the structure of the gel and turning it into a liquid.

Pure sodium polyacrylate can be obtained from magic and novelty shops, or science specialty stores. If a teaspoon of this powder is added to an empty cup, a full cup of water can then be poured in, creating an instant gel. You can then turn the cup upside-down and it will not fall out! Magicians sometimes use this powder to make water "disappear."

Polymers – Experiment # 8:

FUN WITH CORN STARCH

Objective: To observe the behavior of a mixture of corn starch and water.

Materials:

- 16 oz box of corn starch
- Bowl
- Large spoon

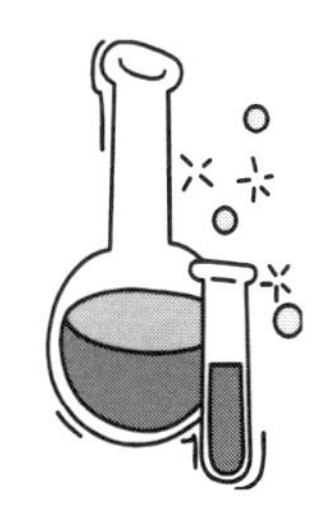

Safety Precautions: None

Procedure:

1. Add the box of corn starch and 300 mL of water to a large bowl and mix thoroughly.
2. Grab a handful of the substance. It will be very runny.
3. Grab another handful, and then squeeze tightly. It will become very hard! When you stop squeezing, what happens?
4. Attempt to poke the mixture in the bowl very quickly with your finger. It will bounce off.
5. Now poke the mixture very slowly with your finger. You can easily touch the bottom of the bowl.
6. Devise your own experiments with this fascinating substance.

Example: The corn starch is an example of a polymer, with many molecules of starch ($C_6H_{10}O_5$) linked together to form clusters. The granules of starch readily absorb water, making starch an excellent thickener for making gravy and sauces. The starch and water together form an interesting example of a non-Newtonian fluid (see Polymer Experiment # 5). Unlike most fluids, the viscosity of this mixture is affected by pressure. As the substance is squeezed it becomes more viscous (less runny). As the pressure is decreased, it becomes less viscous (more runny).

Quicksand, which is a mixture of sand and water, is subject to the same principles as the cornstarch and water. If trapped in quicksand, the more you thrash around, the tighter its hold on you, since sudden movements increase its viscosity. Trying to escape quickly from quicksand is impossible. But if you move slowly through quicksand, its viscosity decreases, making it easier to escape.

Polymers – Experiment # 9:
MAKING A SUPER BALL

Objective: To make a bouncy super ball.

Materials:
- Sodium silicate solution (available in drug or hardware stores, in the paint section)
- Ethyl alcohol (available in hardware and drug stores, as denaturated alcohol)
- Measuring cups
- Plastic sandwich bag

Safety Precautions: Sodium silicate is toxic if ingested, and irritating to the skin and eyes. Wear safety goggles. Denatured alcohol may be fatal if ingested, and is flammable. Keep away from open flames.

Procedure:
1. Add 20 mL of sodium silicate solution and 10 mL of ethyl alcohol to a sandwich bag. Seal the bag.
2. Tip the bag so the liquids accumulate in one corner, and mold them into a ball through the bag.
3. Remove the ball and rinse with water, continuing to shape it into a ball.
4. You may bounce the ball, but it will shatter if bounced too hard. If the ball shatters, squeeze the pieces together with your hand under running water.
5. Place the ball on a flat surface. Observe after a few minutes. What happens?
6. Store in a Ziploc bag.

Explanation: The ethyl alcohol and sodium silicate bond together to form a silicone polymer, which has the properties of rubber. Cross-linking occurs between the silicate particles of the sodium silicate and the ethyl group (– CH_2CH_3) of ethyl alcohol. As a result, long chains of the silicone polymer are formed, which has rubberlike properties. As a result, the ball you made will bounce. The ball appears to be a solid, but is actually a very viscous (slow-moving) liquid. It will slowly flow to assume the shape of its container. It will flatten out into a puddle if left on the table.

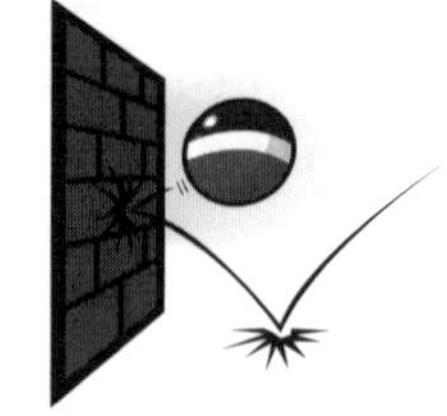

Polymers – Experiment # 10:

LAUNDRY STARCH SLIME

Objective: To make slime that has the consistency of "snot."

Materials:

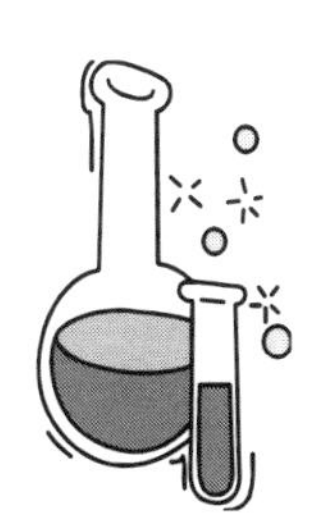

- Elmer's white glue
- Liquid laundry starch
- Food coloring
- Disposable cups
- Popsicle stick for stirring

Safety Precautions: The slime formed from this experiment is toxic if ingested. Wear safety goggles.

Procedure:

1. In a disposable cup, add equal volumes of glue and liquid starch.
2. Add a few drops of food coloring, if desired.
3. Stir for 5-10 minutes.
4. Allow to sit for another 5-10 minutes.
5. Remove the slime with your fingers. It will have the consistency of "snot."
6. Store in a Ziploc bag.

Explanation: If you read the label on a container of liquid laundry starch, you will find that it contains corn starch and borax, among other things. Starch is an excellent thickener, due to its ability to absorb water. As we saw in an earlier experiment (Polymers – Experiment #5), borax causes crosslinking between the polyvinyl acetate molecules that make up Elmer's glue. As a result, a viscous substance is produced, which creates an excellent variation of slime.

CHAPTER 10
LIGHT

Light is a type of electromagnetic radiation, which radiates outward in all directions from its source at the incredible speed of 186,000 miles per second (300,000,000 meters per second) in a vacuum. Other types of electromagnetic radiation are radio waves, microwaves, infrared, ultraviolet, X-rays, and gamma rays. Light travels in waves, as does all electromagnetic radiation. Light is the only part of the electromagnetic spectrum we can see.

The experiments in this chapter should help to shed some visible electromagnetic radiation on this illuminating topic . . .

Light – Experiment #1:

PRODUCING COLORED FLAMES

Objective: To produce different-colored flames using various salts.

Materials:

- Table salt (sodium chloride)
- No-Salt salt substitute (potassium chloride)
- Copper(II) sulfate (available as a root killer in hardware stores)
- Boric acid (available in drug stores)
- Calcium chloride (available as an ice melter)
- Cotton swabs
- Propane-burning camping stove, or candle
- Tongs
- Scissors

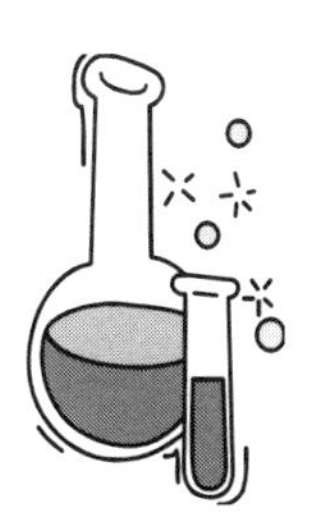

Safety Precautions: Copper(II) sulfate, boric acid, and calcium chloride are toxic if ingested. Exercise caution around flames. Wear safety goggles.

Procedure:

1. Cut several cotton swabs in half with a pair of scissors.
2. Hold one end of the cotton swab with tongs, and dip the other end in water. Then dip the swab in one of the above salts to be tested.
3. Pass the swab through the flame of a gas stove or a candle and note the flame color. (This works best in a dark room. Best results are obtained with a blue flame, which is produced by natural gas or propane. A candle can also be used, but the yellow flame may obscure some of the colors.)
4. After testing, drop the swab into a disposable cup of water.
5. Repeat with each of the other substances listed above. Note the flame color of each.

Explanation: Many different colors can be produced during flame testing, depending on the substance being tested. This is an excellent way to identify various substances. It is generally the metal cation (positively charged ion) that produces the flame color. The sodium ion in sodium chloride will produce a yellow-orange flame color and the calcium ion will produce an orange flame color. The potassium ion will produce a violet flame color, and the copper (II) ion will produce a green flame color. Boric acid will produce a lighter green flame color.

If you have access to a lithium compound, it produces an intense bright red flame color. Strontium compounds will produce a beautiful brick red flame color.

These different colors are due to differences in the wavelengths of the visible light being released during flame testing. As the substances are heated, their electrons absorb energy, which causes them to become excited and jump up to a higher energy level. However, they immediately fall back down to their ground state, releasing this previously absorbed energy as visible light. The energy released will often be of a different wavelength than that absorbed. If the wavelength of the energy emitted is in the visible light spectrum, it can be observed with the naked eye.

The different colors produced during fireworks are due to some of the same substances you just tested.

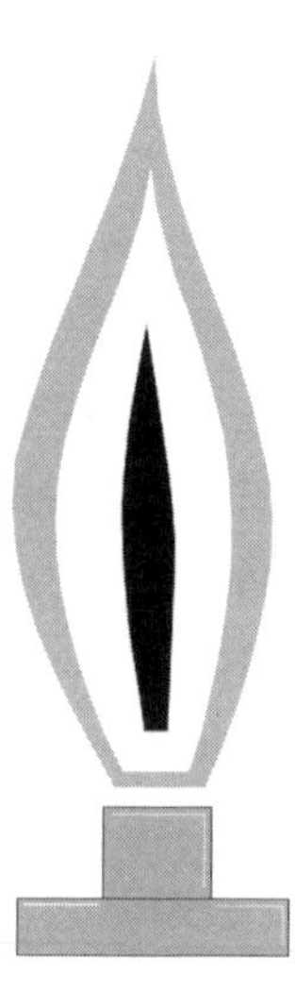

Light – Experiment #2:
"BENDING" A PENCIL

Objective: To demonstrate that light travels slower through water than air.

Materials:
- Plastic sandwich bag
- Sharpened pencil

Safety Precautions: None

Procedure:
1. Fill a sandwich bag halfway with water and then seal.
2. Poke a very sharp pencil through one end of the bag and then out the other end. The bag should not leak.
3. Observe the pencil while in the bag from several different angles. It will appear to be noticeably bent.

Explanation: The plastic sandwich bag does not leak because plastic is a polymer, containing long chains of molecules. The pencil slips in between these chains, which then close back around the pencil, creating a watertight seal.

The pencil appears to be bent because light travels slower through water than air. The water causes the light to be refracted, or bent. Since any object we see is the result of light being reflected from its surface, the optical illusion of the pencil being bent is created.

Light is also refracted when it enters the Earth's atmosphere from space. The atmosphere slows down light waves from the sun. This explains why it gets light in the morning before the sun rises, and also why it remains light for a short time after sunset. The atmosphere bends the light rays, enabling us to see light even though the sun itself is not visible.

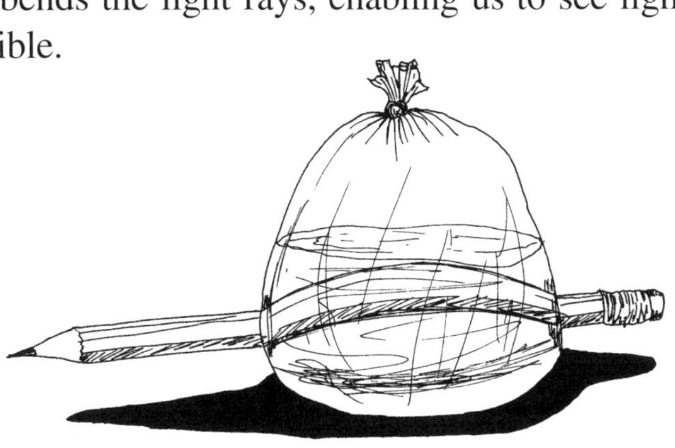

Light – Experiment # 3:
OBSERVING FLUORESCENCE WITH A BLACK LIGHT

Objective: To demonstrate the fluorescence of certain pigments by using a black light.

Materials:
- Black light (available at hardware stores, gift shops, and pet shops)
- Fluorescent markers, crayons, or paints

Safety Precautions: Do not stare directly at the black light; ultraviolet light is harmful to the eyes.

Procedure:
1. Draw or paint a picture using the fluorescent markers, crayons, or paints.
2. Completely darken the room, and observe the picture under the black light.

Explanation: A black light utilizes long wave ultraviolet (UV) energy. Although not as harmful as short wave UV light, which causes sunburns, care should be taken not to stare directly at the black light. UV light is invisible to humans; however, many insects can see UV light. Many commercially available "bug zappers" use UV light to attract insects. The violet light you see when a black light is turned on is actually visible light, which borders on the UV portion of the electromagnetic spectrum.

Fluorescent markers, paints, and crayons contain certain fluorescent pigments that absorb UV light and immediately release this energy as visible light. When UV light is absorbed, electrons within the pigments are excited, which immediately jump up to a higher energy level. As these electrons fall back to ground state, they emit visible light. This emission of visible light causes the fluorescent pigments to glow when viewed under a black light. Fluorescence is defined as the property of absorbing UV light and then emitting visible light.

Light – Experiment # 4:

DO LAUNDRY DETERGENTS REALLY GET YOUR CLOTHES WHITER THAN WHITE?

Objective: To discover why laundry detergents are able to "brighten" your clothing.

Materials:

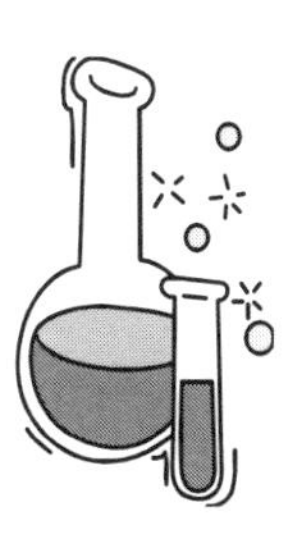

- Black light
- Laundry detergent containers (Check in the trash can of your local laundromat)
- Sample of dry laundry detergent
- Sample of liquid laundry detergent
- Woolite detergent
- Black construction paper
- Articles of white clothing

Safety Precautions: Laundry detergents are toxic if ingested, and are irritating to the skin and eyes. Wear safety goggles. Do not stare directly at the black light; ultraviolet light is harmful to the eyes.

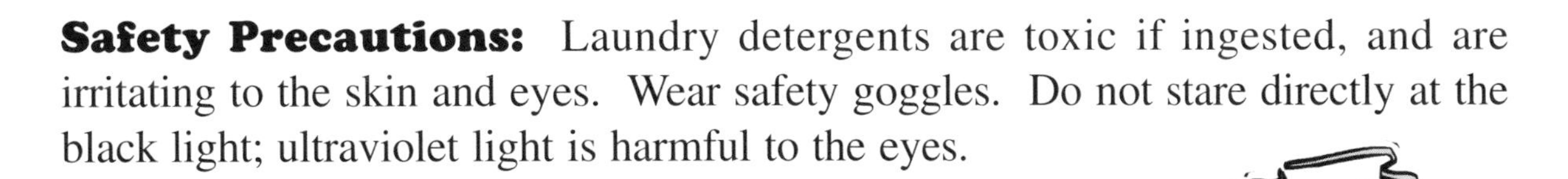

Procedure:

1. In a dark room, observe various laundry detergent containers under the black light. Most of the containers will fluoresce.
2. Sprinkle some dry laundry detergent on a piece of black construction paper and observe under a black light.
3. Repeat with the liquid detergent and then with Woolite. Observe under a black light.
4. Observe articles of white clothing that have been washed in detergent under the black light.

Explanation: Most laundry detergents contain fluorescent pigments that are easily visible under the black light. These pigments give clothing a "whiter than white" appearance. Under the black light, these pigments fluoresce vividly. These pigments have the ability to absorb invisible UV light and release it as visible light, causing a glowing appearance. This same effect is also somewhat visible outdoors under direct sunlight, since sunlight is about 11% UV light. Articles of white clothing also fluoresce vividly under the black light due to residual laundry detergent remaining in the clothing. This demonstrates that most of us use a great deal more laundry detergent than is necessary. Similar pigments are also used in the construction of the laundry detergent containers themselves.

Light – Experiment # 5:

TESTING HOUSEHOLD SUBSTANCES FOR FLUORESCENCE

Objective: To determine which household substances contain fluorescent pigments.

Materials:

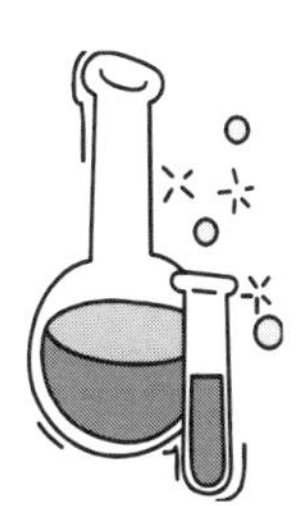

- Black light
- Antifreeze
- Samples of cosmetics
- White copy paper
- Postage stamps
- Fluorescent paper
- Neon-colored clothing
- Brightly colored toys
- Glass jar

Safety Precautions: Antifreeze is toxic if ingested. Read the labels of all household substances before using. Do not stare directly at the black light; ultraviolet light is harmful to the eyes.

Procedure:

1. In a dark room, pour a little antifreeze into a glass jar. Observe under a black light.
2. Observe each of the other substances listed above under a black light.
3. Try to find other substances around the house that will fluoresce under a black light.

Explanation: Many ordinary household substances fluoresce vividly under the black light. The next time you visit the mall, look for black light sensitive objects in gift shops or science specialty stores. You can buy fluorescent posters, candles, clothing, nail polish, hair spray, and jewelry. Any object that fluoresces under the black light contains fluorescent pigments that absorb UV light and emit visible light.

Light – Experiment # 6:

HOW DO SUNSCREENS WORK?

Objective: To determine how sunscreens prevent you from getting a sunburn.

Materials:

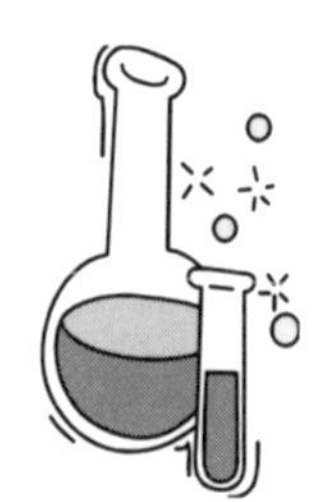

- Black light
- Sunscreen lotion
- Vaseline
- Elmer's glue

Safety Precautions: Do not stare directly at the black light; ultraviolet light is harmful to the eyes.

Procedure:

1. Squirt some sunscreen on a piece of white paper.
2. Observe under the black light.
3. Repeat with Vaseline and Elmer's glue.

Explanation: Sunscreen lotions work by absorbing UV light. As UV light is absorbed, no other light is released; thus it appears to be black under the black light. Vaseline and Elmer's glue have similar properties; however, their use as a sunscreen is not recommended! Experiment with other household substances – anything that appears black under the black light is absorbing the UV light.

Since sunscreens absorb UV light, they protect our skin from harmful UV rays. Sunburns are caused by shorter wave UV rays known as UVB light. The black light is composed of longer, less harmful UV rays, known as UVA light. Sunscreens are capable of absorbing both UVA and UVB light.

Light – Experiment # 7:

HOW WELL DO YOU WASH YOUR HANDS?

Objective: To determine how well you wash your hands by using a black light.

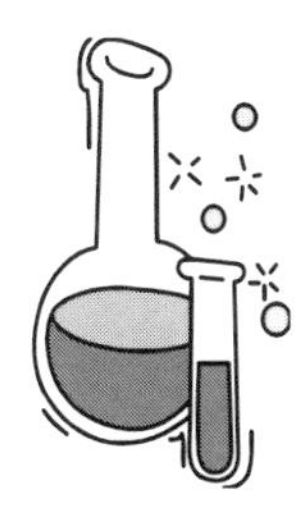

Materials:

- Black light
- Glo Germ lotion (Available from the Glo Germ Company at 800-842-6622 or www.glogerm.com)

Safety Precautions: Do not stare directly at the black light; ultraviolet light is harmful to the eyes.

Procedure:

1. Squirt a small amount of Glo Germ lotion on your hands and rub in thoroughly.
2. In a dark room, observe your hands under the black light. They will glow with a bright orange color.
3. Wash your hands thoroughly with soap and water.
4. Observe your hands again under the black light. Did you miss any areas of your hands when washing?

Explanation: The Glo Germ lotion is a highly fluorescent oil-based fluid that fluoresces vividly under the black light. It can be purchased inexpensively from the Glo Germ Company.

Glo Germ lotion was originally developed as a tool to help train medical personnel in the proper technique of hand washing. Today, it is also widely used in the restaurant industry, as well as other consumer based industries where proper hygiene is essential. It has even been used by the Atomic Energy Commission to train workers in the proper handling of radioactive materials.

As you have probably discovered, it is not easy to remove every trace of the Glo Germ lotion from your hands. If the lotion represented germs, all traces left on the hands would contain potential pathogens. The Glo Germ lotion is actually formulated to contain materials that are the same size as a bacterium – about 5 microns in diameter.

Visit the Glo Germ website (listed above) to learn more about this amazing substance. Their website also contains several lesson plans for educators.

Light – Experiment # 8:

LIGHT FROM A LIFESAVER

Objective: To demonstrate how a wintergreen Lifesaver can produce light.

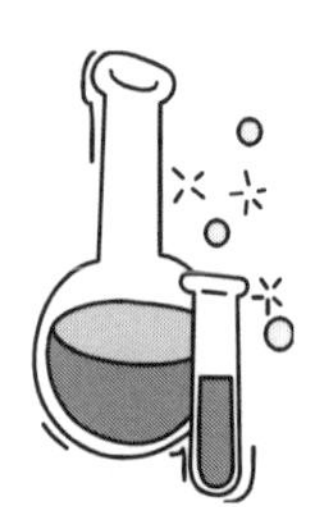

Materials:

- Wintergreen Lifesavers
- Pliers

Safety Precautions: None

Procedure: In a completely dark room, crush a wintergreen Lifesaver with a pair of pliers. Observe.

Explanation: A bluish light can be readily observed when the wintergreen Lifesaver is crushed with a pair of pliers. It is easier to see if the room is completely dark. As the sugar crystals that make up the Lifesaver are broken apart, they release energy through a process known as triboluminescence. This process occurs when energy is released as a result of the mechanical fracture of crystals.

The energy produced in this experiment is similar to lightning, since it is created by an excess of electrons passing through air. When the sugar crystals that make up the Lifesavers are broken, they tend to separate along planes that are oppositely charged. The excess of electrons on the negatively charged plane tries to bridge the gap between the two planes. These electrons excite the electrons within the nitrogen molecules in the air. As the electrons from the nitrogen molecules fall back to ground state, they mostly emit energy in the UV range. The wintergreen flavoring in the Lifesaver is fluorescent, therefore it absorbs this UV light and emits visible blue light that we can see.

Similar effects can be observed if rock candy or sugar cubes are crushed in the dark, or a piece of Scotch tape is quickly removed from the roll. However, the light emitted from these objects is usually less intense due to the lack of the fluorescent wintergreen flavoring.

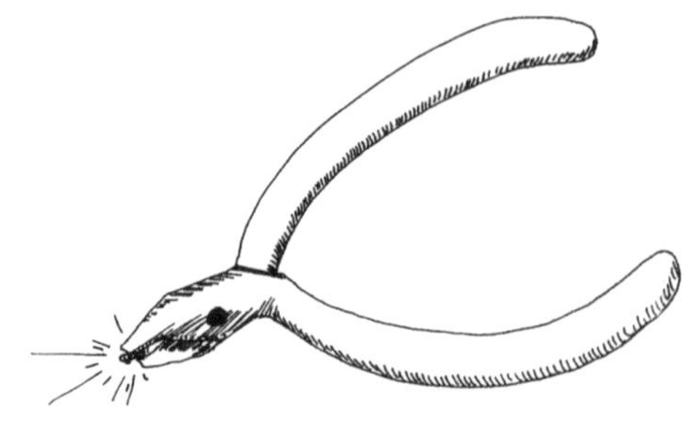

Notes:

Notes: